Rooted

Also by Lyanda Lynn Haupt

Mozart's Starling

Urban Bestiary

Crow Planet

Pilgrim on the Great Bird Continent

Rare Encounters with Ordinary Birds

Rooted

Life at the Crossroads of Science, Nature, and Spirit

LYANDA LYNN HAUPT

Illustrations by Helen Nicholson

LITTLE, BROWN SPARK

New York Boston London

Little, Brown Spark
Hachette Book Group
1290 Avenue of the Americas, New York, NY 10104
littlebrownspark.com

First Edition: May 2021

Little, Brown Spark is an imprint of Little, Brown and Company, a division
of Hachette Book Group, Inc. The Little, Brown Spark name and logo are
trademarks of Hachette Book Group, Inc.

The publisher is not responsible for websites (or their content) that are not
owned by the publisher.

The Hachette Speakers Bureau provides a wide range of authors for
speaking events. To find out more, go to hachettespeakersbureau.com or call
(866) 376-6591.

Copyright acknowledgments appear on page 228.

Printing 4, 2022

ISBN 978-0-316-42648-0
LCCN 2020951183

LSC-C

Printed in the United States of America

To David and David—
with gratitude for more support than
I could ever deserve

Contents

Frog Church

A Rooted Invocation 5

The Tenets of Rootedness 23

Listen

The Wild Summons of the Wolf Path 29

The Wildflower Task

The "I Am Here" of Hope

Pilgrim on the Wolf Path

Listening to the Wolf

A Wild Summons

Shed

Bare Feet, Holy Ground 45

What We Leave Behind

The Megalithic Measure

Walking Clay

Intelligent Feet

Human Paws and the Dishonorable Toe

Homecoming

CONTENTS

Wander
A Feral Cartography 61

Skeletons

Dangerous Business

Wired to Wander

The Wandering Mind

Songlines / Selflines

Wandering for Beginners

Immerse
Staying Dirty During a Forest Bath 77

Forest Prescribed

Forest Inspirited / Forest Commodified

The Replaceable Wild

The Calm Anxiety of Creation

Staying Dirty During a Forest Bath

Alone
The Essential Complexity of Solitude 93

A Culture of Unsolitude

Safe in the Woods

The Primal Sanity of Solitude

The Solitary Brain

Flourishing

Return

Unsee
Return to a Fruitful Darkness 111

Footsteps in the Dark

When Darkness Became Evil

CONTENTS

The Sheltering Darkness

Darkness Lost

Human Animals in the Dark

Life at the Speed of Dark

Relate
An Infinity of Animal Intelligences   129

The Doe

The Orca

Beyond Anthropomorphism

A Multiplicity of Intelligences

Coming into Animal Presence

Shapeshifting

Lifting Up

Speak
The Abracadabra of Earth   149

True Naming

Excavating True Names

Speaking of Others

Behold

Grow
Trees and the Rings of Belonging   165

The Grieving Trees

Truth and Trees

Growth Rings

Communion and Kith

Forest Risk, Forest Attunement

Weirdos Among the Trees

CONTENTS

Create
The Art of Earth Activism 185

The One Thing and the Ten Thousand Things

Create-Creature-Creation

The Creating Wind

The Science of Creatures Creating

The Creative Ecology of Home Economics

Name Your Charism

Spiral
A Wild Return 203

Skin and Soil

The List

Dress for the Occasion

An Earthen Immortality

Acknowledgments   219

Selected Bibliography   221

Copyright Acknowledgments   228

Rooted

Our hands imbibe like roots,
so I place them on what is beautiful in this world.

—Francis of Assisi

Frog Church

A Rooted Invocation

When I was in fourth grade, my mother put a copy of Saint Thérèse of Lisieux's diary, *The Story of a Soul,* in my Easter basket, right alongside the marshmallow bunny. What was she thinking? I devoured the words of this fervent, neurotic, ecstatic young woman who saw the divine in the way a chicken cares for its young, and quickly took up the study of other nature mystics who innately apprehended the graced interconnection of life—Julian of Norwich, Hildegard of Bingen, Francis of Assisi. They shed their shoes, followed bear tracks, declared the moon a sister, spoke with sparrows, ground forest nettles into healing salves, bowed before trees, baked bread in clay ovens, and called all of it holy. From them I learned that humans can be conversant with the earth and the sacred in strange, imaginative, wild ways. *In any way we want.*

This is not a book about religion, so if you are religious, or not religious, don't worry. But my adult relationship to nature is rooted in my childhood experience. I know I'm in the minority when I say that I loved being raised Catholic. People talk about being "recovering Catholics," and "Catholic guilt," for good reason.

But when young, I knew nothing of religious politics, or of a male-only priesthood (I thought our priest just happened to be a man), or of the yet-to-come sexual-abuse scandals, or reproductive freedom, or marriage equality. Guilt was not emphasized in my church or my home. I knew only that my developing mind responded to the religion's playground of imagery both mystical and untamed.

Other childhood friends were not Catholic, and when we ventured to speak of family Sundays I would sit silently wondering. Where were their angels, their candles, their magical water from a giant font that they were allowed to take home in baby-food jars and sprinkle on their heads to keep them safe during sleep? Where were their saints running wild in the desert, in the wilderness, communing with wolves in hidden mountain caves? Where was their young Blessed Christina the Astonishing, who was believed to be dead but, at her own funeral, sat up in her coffin to complain of the priests' stench and flew first to the rafters, then to a tree, where she lived among the birds, refusing to come down ever again? Where was their blood? Next to such feral glories all other churches (or no church at all) seemed impoverished.

I didn't even mind the confessions we were required to make each month, though they struck fear in my friends and my little sister. It was certainly a bit overwhelming, entering a darkened room all alone, facing a priest behind a wood-carved grate, and being made to recite my supposed sins. No matter what the catechism taught, I'd never believed that a man was needed to intercede for me on behalf of God, whatever God might be. But I believed in the power of sacrament, in very much the way I do today—not as a Catholic but as a human open to the truth that something can be made sacred by the attention we grant it.

And so one day in fifth grade when it was my turn, I stepped into the narrow confessional and knelt. The room smelled faintly of urine, for it was rumored that Mikey Roberts had been so scared at his own confession the month previous, he'd peed his new corduroy pants. I ignored this and readied myself for my impending purity.

"Bless me, Father, for I have sinned," I whispered as I made the sign of the cross. I could see Father Sandergeld's large nose from my side of the grate—red from his daily motorcycle rides before the advent of sunscreen, but also, we all assumed, from the drink that kept him company most nights. He leaned his head into his hand, pretending, as was the custom, not to see who was there through the insufficient screen.

"Yes, my child?"

"I am often mad at my mother." Father sighed. What sins does a fifth grader really have? He must have heard this a million times.

"Be patient with your mother. She loves you very much."

"I am mean to my sister."

"She is younger than you." I knew he could see me—how else would he know I was the eldest Haupt sister? "It is up to you to set a good example. For your penance, say ten Hail Marys."

Things were going my way. I *loved* the Hail Mary. To me, Mary was a mother-goddess who appeared to orphan children in poor countries, who in her guise as the Lady of Guadalupe, threw tropical roses upon the impoverished cloak of Juan Diego though it was winter, whose own cloak shimmered with stars.

She protected all of nature. I longed for her to appear to me and often wept that she did not. Praying to her came naturally.

"I will say twenty!" I announced fervently, causing Father to sigh again.

"Ten will suffice."

"There is one more thing."

I loved having a secret, but I yearned to tell *someone,* and I knew of no one else who would not intrude once they heard. Still, I was a little frightened. Maybe this really was a sin. *One Holy Catholic Church,* the creed proclaimed, after all. Perhaps Father Sandergeld would be angry or tell my mother or demand that I stop altogether.

I summoned my courage.

"I have another church." There was a long pause and I sensed this was new territory for the priest. He didn't seem to know what to say. But finally he lifted his head.

"Lyanda," he ventured, giving up the pretense of anonymity altogether, "what is this church?"

I told him everything in one long breath, then gasped, "Do I have to tell my mom? Or anyone?"

Father put his head back in his hand, from practice or exasperation I could not tell. "God knows all you do." I'd already learned this in catechism class, and nodded in the darkness. "I suppose that is enough," he mused. "But you'd better add an Our Father to your penance."

I emerged joyous, radiant as an angel, the sacrament in full force. I did not give a backward glance to my suffering fifth-grade colleagues, who were dreading their turn in the confessional. When I got home I went straight to my other church.

Behind the house where I grew up there was a vacant lot covered in tall brown grasses, and beyond that there was a wooded canyon with narrow trails among Douglas firs, western red-cedars, bigleaf maples, and all manner of ferns and mosses, leading down to a stream called, like many streams with a forgotten history in this country, Mill Creek. Overinfluenced by *Anne of Green Gables,* I did not call the creek by its map-name, but by what I believed in my romantic young heart to be its soul-name. And it was there, on the still edges of the Stream of Sparkling Stones, that I discovered Frog Church.

It was a small and secret church with few requirements for its single human attendee—none of them difficult for an introverted child who carried only an apple and a sketch pad that doubled as a diary. Rules: no shoes; quiet always; drawing and writing allowed if they did not cause disturbance, as wild animal visitations were the most sacred treasure.

The longer I stayed, and the quieter I was, the more creatures would wander through. Squirrels and robins and chickadees. Once a hummingbird landed on my head. Once a garter snake slid by and brushed my heel with soft skin. Once a striped skunk waddled through, followed by four perfect baby skunks. None of them even glanced at me.

But frogs were special. Looking back I suppose it is because they were the only animals other than insects that I could easily catch. But there was something, too, about their odd fluidity. Wandering between worlds. Egg, tadpole, froglet, frog. Fins to feet. Water to earth and back again, over and over.

The rules at the creek were entirely different when I was scrambling there with friends—mostly the neighborhood boys and my sister. At those times the goal was to first spy, and then to capture all the frogs we could. This was not Frog Church at all. We splashed and screamed and got so muddy it was impossible to tell us apart.

But I always looked forward to slipping away to visit the frogs on my own. I learned that if you approach quietly and do not cast a shadow upon them, frogs will not leave their stones or the quiet stream edge, where they rest with their eyes showing just above and their bodies just below the clear waterline. And I learned further that if you are quiet and slower than slow, you do not have to "catch" a frog at all—you can just slide your hand beneath one and lift it up without inciting any hint of fear or effort to escape. I learned that if you do these things, a frog will be calm and settle in your palm, and that, once settled, it will be happy to stay, and that the frog's willingness in this regard filled me with serenity, and with joy, and with an unfolding sense of nature's irrepressible interconnection.

I learned that it was hard to hold more than one frog calmly in my hands but if I would lie on my back close to the stream and move the frogs from my hands to my bare tummy they would stay—soft cold feet and strange round breathing. In this way I could reach the stream with one arm and gather more frogs for my congregation. When a sufficient number was assembled (three or four frogs and me), we would all (I chose to believe) hum and pray together, belly to belly, accompanied by birdsong and the whispers of trees overhead, and thoughts—sweet, simple, and blessedly few.

Eventually I would hear my mother. "Lyaaaaaandaaaaa!" It was dinnertime, and she was calling me home. My camp at the base of the canyon was far from our house's back door—it was plausible that I couldn't hear my mom, and I made use of this likelihood when it suited me. But now I suddenly realized how hungry I was. "Go in peace," I whispered to the frogs, poking one on the behind to wake her up. The movement stirred the others and suddenly all of them were in the stream, and gone. Frog Church for the day was over.

For thousands of years we have struggled with the human condition under the assumption that this condition, whatever its faults, would continue. *Now and forever,* the creed of my childhood faith reads, *now and forever.* But now, for the first time in human history, we are living at a juncture where the twin realities of climate crisis and habitat destruction are so far-reaching that the basic web of biological connections required to support life on earth are swiftly breaking down. People are rightfully experiencing unprecedented anxiety and despair as we contemplate our place on this planet, and what is asked of us.

As perilous and complex as these times are, we are armed with a rare trio of tools that offers a rooted way forward: the joining of nature, spirit, and a uniquely modern science. The innate connection between humans and the natural world is coming to the fore in a new way, as academic research rises in support of truths that poets, writers, mystics, artists, naturalists, earth-based religions, and Indigenous cultures have always proclaimed. Multiple studies appear in the most respected journals, including *Science, Nature,* and *The Journal of the National Academy of Sciences,* detailing evidence that: trees are able to communicate with one another, both through the motions of their branches, spreading chemical messages aboveground, and through a web of intertwined roots and fungal mycelia belowground; all animals have distinct and complex consciousnesses and intelligences and languages beyond what we have ever scientifically acknowledged; humans are more creative, physically hale, and less depressed after walking in a forest; wandering barefoot upon the earth improves podiatric health and increases the physical intelligence of our whole being. Our bodies, minds, and spirits stand in ancient communion with the soil.

In the genre of philosophical nature writing, academic science is sometimes suspect. Philosopher and wilderness guide Jack Turner writes that "science disembeds individual context, people, flora, fauna." The late theologian, eco-cultural historian, and self-described "geologian" Thomas Berry echoes this sentiment:

> While we have more scientific knowledge of the universe than any people ever had, it is not the type of knowledge that leads to an intimate presence within a meaningful universe.... The difficulty is that with the rise of the modern sciences we began to think of the universe as a collection of objects rather than as a communion of subjects.

Many friends report that they don't need the mechanistic language or quantifiable evidence of a scientific paper to tell them that animals are intelligent or that they feel better in body and spirit after a woodland ramble. And it's true. When we walk mindfully in the natural world—attuned to the voices of the birds, the alternate unfurling and dormancy of plant life through the seasons, the tracks of the coyotes who wander from the green space to our urban backyard—a great deal of truth about the interrelationship between humans, plants, animals, and the land is *directly* revealed to us. This attunement is our most authentic, most innate way of learning and knowing.

Yet just as many friends tell me that the recent science has made them more comfortable talking—or even thinking—about such things. In countless ways, modern culture and urban habits contrive to separate us from our indwelling earthen intelligence. Journalist Florence Williams reports in her book *The Nature Fix* that North Americans spend over 93 percent of their waking hours indoors or in cars (and the other 7 percent is spent walking *between* buildings and cars). And while there is an extraordinary exuberance and diversity of wild plant and animal life dwelling in our midst, still the urban environment is inhospitable to the majority of species on our shared planet. Regular—or *any*—experience of deep wilderness is missing from most of our modern lives. Without such contact, our radiant mental and physical intelligences are being diminished. Williams writes:

Thanks to a confluence of demographics and technology, we've pivoted further away from nature than any generation before us. At the same time, we're increasingly burdened by chronic ailments made worse by time spent indoors, from myopia and vitamin D deficiency to obesity, depression, loneliness and anxiety.

Hildegard of Bingen was a Rhenish mystic, musician, composer, prophet, herbalist, pharmacist, and healer (at certain times she would have been called a witch, in spite of her Benedictine monastic garb, and these days she is heralded as such in certain circles as this word is reclaimed). Hildegard would not have been surprised by the effects of our separation from nature: in the 1100s she referred to people deprived of contact with the "lush greenness" of the landscape as "shriveled and wilted."

Doubters about the great aliveness of the wild earth — people who question their inner knowing, or those who simply have not experienced a personal connection with the more-than-human-world — may be brought, through the certainty of scientific inquiry, to a new way of seeing. So many people have felt somewhere in their core a spiritual connection to nature but have not had the language to ponder it deeply in our evidence-heavy over-culture; the new science gives them this language, and the essential courage, to trust the intuition they have always possessed. And for those who would deny that the earth is alive, life-sustaining, *and* at a perilous anthropogenic tipping point? Well, science is giving them no place to hide.

The modern science of nature is significant for many other reasons, beyond the obvious setting of conservation priorities and actions. Foremost in my mind being the fact that it is beautiful. Its wondrous mathematical synchronicities, the specifics of its chemical analyses, the complexity of its physics are beyond both the practical and intuitive knowledge of most lay naturalists (or mystics), no matter how seasoned. When mingled with the wildness of the natural world and the creativity of the human mind, good science reveals its center, its story, its deeper teaching. The science has its own poetic force.

Yet no matter how significant the research, I believe that spirit—the response of our hearts and imaginations to the whole of life, so often beyond traditional rationality—is required to fully animate the new science. The quantified results of scientific work, and the stringent lexiconic language in which they are reported, awaken and sing through the brightness of our ensouled stories, unfolding in concert with nature. The poetry of earthen life cannot reach its fullness on a computer screen, or even in the synapses of our magnificent intellect. Our hearts are formed of a wilder clay.

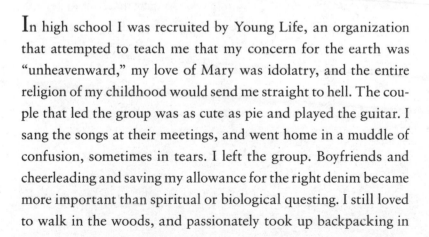

In high school I was recruited by Young Life, an organization that attempted to teach me that my concern for the earth was "unheavenward," my love of Mary was idolatry, and the entire religion of my childhood would send me straight to hell. The couple that led the group was as cute as pie and played the guitar. I sang the songs at their meetings, and went home in a muddle of confusion, sometimes in tears. I left the group. Boyfriends and cheerleading and saving my allowance for the right denim became more important than spiritual or biological questing. I still loved to walk in the woods, and passionately took up backpacking in wilder places, though by this time my town had "improved" the trail along the Stream of Shining Stones—it was buried now under an asphalt path and hosted a paved trail's attendant joggers. I couldn't find a frog near my house if I tried. My saints, too, seemed to fall silent.

But as a philosophy major in college, alongside the ancient Greeks and the modern existentialists, I studied world religions and wisdom, especially Buddhist thought. One night, while poring over a book of Japanese art, I became enraptured by a worn camphorwood statue of a quiet-eyed deity, resting their head on their fingertips, looking more peaceful than anything I had ever seen constructed by human hands. I was shocked to see the slender, androgynous entity referred to as "she" in the description. Nearly all of the deities I was familiar with in the ranging Japanese canon were male.* This was Kuan Yin, or Kannon in Japanese—the bodhisattva of compassion. I fell in love with the statue, and literally ran across the grassy quad to my Japanese professor's office, book in hand. It was dark and nearly midnight, but as expected, Takemoto-sensei sat quietly in his office, sipping hot green tea. He poured some for me and just smiled and nodded as I pointed at the caption. Almost immediately, he started to help me make arrangements for a year of independent study in Japan.

I traveled to Japan, lived in Kyoto, walked everywhere, haunted the temples, practiced tea ceremony obsessively, sat *zazen,* fell in love with a monk who vowed to marry me (even my

---

* Later, while studying in Japan, I learned that there were more female deities than I had realized.

romantic young self could see that would never work, but we had some fun), and helped translate Japanese texts into English at the local university in exchange for instruction in Buddhist philosophy and literature. Whenever I could, I took the train to Nara to sit at the feet of the beautiful image of Kuan Yin, tearful over her beauty (and slightly afraid to leave her temple with all of Nara's famous and aggressive tourist-fed deer waiting outside to nibble my elbows). Here was the compassionate Mary of my youth in a different form. I'd missed her. And here, far from home, the voices of the mystics returned with the force of grace and the love of family. I saw more clearly than ever that I didn't need a particular organized religion to live in a manner that embraced their wild teachings.

Meanwhile, the forest wandering of my childhood had grown into a zealous love of biological science, natural history, and birdwatching. I felt gratefully rooted in the old tradition where science, nature, philosophy, aesthetics, and religion were not different academic trajectories but all of a piece—"natural philosophy" in Victorian parlance. Under this generous intellectual umbrella, Rachel Carson, Charles Darwin, Hildegard of Bingen, and Georgia O'Keeffe were among my many mentors.

Later, studying environmental ethics in a graduate school philosophy department where my vision of natural philosophy was accepted as normal felt like heaven. I studied trees and neotropical migrant birds and wolves and live-birthing frogs (then on the edge of extinction, now vanished from earth forever) and their relationship to—well, to everything. To radical environmentalism and Japanese poetry and transpersonal ecology and the earth-based mysticism that had never ever left me. I spent all my extra time wandering the natural area near my school in Colorado with bin-oculars around my neck and a silly grin on my face.

I returned home to the Pacific Northwest to write my master's thesis on radical environmental activism, the first draft scribbled by pencil in notebooks on long solo backpack trips among the mountains and rivers of the moody green Olympics. I took the train back to Colorado to defend my thesis. And though I loved ecological philosophy, I ran away from academia the second I finished my degree, wanting to engage the subject in a more direct way — which is turning out to be a lifelong project.*

It has been decades since my family moved from my child-hood home near the frog-creek, and now I have my own home and family in Seattle. When the fraught name God comes up in conversation or reading, I always remind myself that whatever the source or language used, we are at root on common ground — invoking the graced, unnamable source of life, the sacredness that cradles and infuses all of creation, on earth and beyond. I know that prayer is the lifting of our hearts, our thoughts, and even our bodies in conversation, or contempla-tion, or remembrance, or supplication, or gratitude within this whole, requiring no dogma, only openness. Hildegard coun-seled, "To be alive is to give praise." The works of the Western mystics and Buddhist philosophers and Emily Dickinson sit on my bedside table and ride in my rucksack, along with well-worn

---

* Of course, many writers find such engagement wonderfully within academia — Robin Wall Kimmerer, Kathleen Dean Moore, David Abram, Robert Macfarlane, and many others come immediately to mind, but I knew it was not my personal path.

field guides to birds and medicinal plants. And my feet are so dirty from wandering barefoot in the woods that I fear they will never be clean.

In this time of planetary crisis, overwhelm is common. What to do? There is so much. Too much. No single human can work to save the orcas and protect the Amazon and organize anti-fracking protests and write poetry that inspires others to act and pray in a hermit's dwelling for transformation and get dinner on the table. How easy it is to feel paralyzed by obligations. How easy it is to feel lost and insignificant and unable to know what is best, to feel adrift while yearning for purpose.

Rootedness is a way of being in concert with the wilderness — and wildness — that sustains humans and all of life. The rooted pathways offered here are not meant as a definitive list but as waymarkers and fortification for all of us seeking our unique, bewildering, awkward way through the essential question of *how to live* on our broken, imperiled, beloved earth. It is the question Thoreau asked. The one that Mary Oliver, who passed just before I wrote these words, has perhaps framed most beautifully: *Tell me, what is it you plan to do with your one wild and precious life?*

We all come to know our own lessons as they spring from the study of our households, our woodlands, our watching, our footprints, the trails of kindred wild ones that cross our paths. Our intelligent feet, our making hands, our listening ears.

The word *rooted*'s own root is the Latin *radix,* the center from which all things germinate and arise. The *radix* is the *radical*—the intrinsic, organic, fervent heart of being and action. Rooted lives are radically intertwined with the vitality of the planet. In a time that evokes fear and paralysis, rooted ways of being-within-nature assure us that we are grounded in the natural world. Our bodies, our thoughts, our minds, our spirits are affected by the whole of the earthen community, and affect this whole in return. This is both a mystical sensibility and a scientific fact. It is an awareness that makes us tingle with its responsibility, its beauty, its poetry. It makes our lives our most foundational form of activism. It means everything we do matters, and matters wondrously.

Amphibious, we wander at the singular, radical intersection of science, nature, and spirit. Here resides a multifaceted understanding of the interdependence of earthly life and the engaged activism that such an understanding inspires and requires. Here are the interwoven pathways of inward, wild stillness and outward, feral action. At this crossroads there is intelligence, and sacredness, and wildness, and grace. There is clear-sighted hope in a time of despair. Rooted ways embolden us to remember that with our complex minds we can feel—and live—more than one thing simultaneously. Anxiety, difficulty, fear, despair. Yes. Beauty, connectedness, possibility, love. *Yes.*

The Tenets of
Rootedness

S everal fundamental beliefs underlie all that grows within
these pages. These tenets are grounded in millennia of
human relations to the wild earth that have come to us across
cultures through writings, art, science, and storytellings. They
have come, too, through what I have found to be true in my own
years of wanderings and wonderings as I ponder how to live with
grace upon a changing earth. Many of the tenets of rootedness
involve words and concepts that have different connotations for
various people and traditions. I offer here a short field guide to
the particular meanings of significant words and ideas as they
are conceived in this project.

Ecology and Mysticism: Rooted mysticism is the apprehension
of life's radical interconnection. Mystical insight and ecological science are mutually reinforcing—both refer to an
underlying unity that dissolves the notion of an absolutely
separate individual. Vietnamese Buddhist monk and Zen
master Thich Nhat Hanh conceived the word *interbeing* in

English to describe the interdependence of all life and phenomena. Walking within this ecological/mystical perspective, a wanderer who finds a feather on a woodland pathway sees not only the bird it once clothed but the forest that sustained the bird, the clouds, the rain, the sun, the seeding darkness. She sees the stars. She sees that her life and that of the feather-bird are indissolubly connected.*

Everyday Animism. We walk within an infinity of other-than-human consciousnesses. Many of these are not recognizable by typical human measures—skeleton, nervous system, brain, legs, or intelligence that looks anything like our own. Yet all of the natural world is inspirited in a manner perfectly befitting *itself*. There is no "mere" matter. All ways of being, from hominid to dandelion to dragonfly to cedar tree, possess a kind of aliveness. One need not be steeped in a particular animistic tradition to live under the influence of such knowing.

Poetry and Science Intermingle. Science speaks from a sphere of rationality and quantification that does not always seem to sing to matters of spirit. Yet within the stringent language and the reliance on statistical significance, good science enhances poetic wonder in the face of a complex universe,

---

* In these pages I refer to mysticism mainly as a mind-set, rather than mystical *experience* as a passing noetic state (though a mystical stance toward the world can certainly invite such experiences). For more on the latter, see William James's *Varieties of Religious Experience* and Carl von Essen's *Ecomysticism*.

bringing the magnificence of the unseen to life. The real issue is that too often science is accepted as the sole arbiter of validity. As Wade Davis wrote in *The Wayfinders,* "Science is only one way of knowing and its purpose is not to generate absolute truths but rather to inspire better and better ways of thinking about phenomena." Poetry, science, story, art, all bring depth and knowing to one another—all mingle as co-expressions of a wild earth.

Truth and Fact Are Not Synonyms. Science gives us the essential realm of fact, more significant than ever in this time of health and climate crises. Yet a poem, a myth, a moment of apprehension may be entirely true while not being scientifically factual.

Mystery. The earth, the creatures, the rivers, the cosmos, all possess inspirited secrets that even the vast excavation of science cannot reveal. It is hubristic to believe that it is only a matter of time and effort before human investigation will expose all that can be known of the natural world. We walk, always, within enlivened mystery.

Kindred, All. The earth and all that dwell herein—the rooted, feathered, furred, scaled—are relations, our dear kindred. Kin is from the Old English, *of the same kind*. Together, we are made of the fine things: clay, salt, water, stardust. We live in a wild communion of absolute mutual dependence, connected in our ancestry and our continuation.

Kith. "Kithship" is intimacy with the landscape in which one dwells and is entangled. In modern English—even in England, where the expression originated—"kith and kin" are mistakenly conflated into one meaning: our relatives, those who are close to us. But the reason the archaic phrase was formed around two different words is that they *are* in fact different. Jay Griffiths points this out in her book, *Kith* (retitled *A Country Called Childhood* for an American audience not trusted to know the word *kith* at all). Where kin are relations of *kind*, kith is relationship based on *knowledge* of place—the close landscape, "one's square mile," as Griffiths writes, where each tree and neighbor and robin and fox and stone is known, not by map or guide but by heart. Kith is intimacy with a place, its landmarks, its fragrance, the habits of its wildlings. Kithship enlivens kinship.

Reciprocity. The earth cradles our human life, providing shelter, food, medicine, health in mind and spirit. So often we measure the value of nature—both personally and in terms of conservation priorities—by the extent to which a place or an organism or even a walk in the forest provides these things. Yet as we partake, so we are called to nourish in return. It is said that "if we walk in the woods, we must feed the mosquitoes." A rooted life is one of reciprocity—the tending of a mutually enhancing presence within the natural world. It may or may not involve mosquito bites. Peaceful footsteps, study, attention, attunement, activism, gratitude, protection, prayer, writing, witness. There are an infinity of ways to be responsive within the living earth.

All Is Sacred. The sacrum is the bone at the base of most vertebrates' spinal column. It is the root of the skeletal apparatus that allows us to take shape and form movement. *Sacrum* shares the Latin root *sacer* with the word *sacred*. A recognition of the sacred in all of nature is the source of any movement toward reciprocity—inner and outer. It hallows our life and work.

Enchantment and Wonder. Moments of enchantment and wonder are typically portrayed as experiences that fall upon us, when in truth they most often require a cultivation of openness to their visitations. The wondr*ous* and the enchant-*ing* are a given—everywhere about us. But wonder derives from the Old English *wundrian*—to be astonished by the presence of the wondrous. So too with *enchantment,* which arises from *chant* via the Old Northern French *cant*—to sing the sacred and (the *en-* part) to be changed by the song we have heard. Wonder and enchantment require us to disengage from culturally constructed norms of rationality for adult humans and *allow* ourselves to be affected by the astonishing world that enfolds us always. It is essential to a rooted life, as Rachel Carson wrote in her essay on the subject: wonder is "an unfailing antidote" against boredom and disenchantment throughout life, against the "sterile preoccupation with things that are artificial, the alienation from the sources of our strength."

Creativity and the Great Work. No matter how many times we hear it repeated as a New Age platitude, it remains true—we each have singular gifts and offerings. This is not about one grand preordained purpose. No, it is about knowing our unique gifts and strengths, and bringing them in service to this troubled and complex time on earth in what Thomas Berry calls our Great Work. I adopt this language along with, simply: creativity, work, activism, art. The joining of our own unique arts to those of the collective whole is the deepest — perhaps the only — hope for the continuation of a wild earth.

Eccentricity. To live rooted on a changing earth is to create a new story. There are very few voices left that speak for wild nature first. It's time to clasp hands (paws, fins, feathers, branches) and know where we stand. The ways of this story will not appear conventional within outworn cultural norms. Our new ways are disruptive. They will look weird. This is good. Let us not care, but enjoy that glimpse in another's eyes that we will find sometimes — the one that says, "You're not crazy. I feel it, too."

Listen

The Wild Summons of the Wolf Path

Sometime last year I began to have recurring dreams of walking on forested pathways, wandering off a large path onto smaller, hidden paths. And in most of the dreams, I was wearing a red hood.

I am a big believer in following dreams and obsessions where they lead, so: I read every academic article I could find about the folktale of Little Red Riding Hood; I checked every illustrated children's book of the story out of the library and put myself to sleep poring over the artists' varied visions. And while I was doing all of that, I knit a pattern called "The Dragon Watcher's Hood," with a bell attached to keep from startling sleepy dragons, which seems sensible. The hood is cozy and warm, and when I wear it, I feel I'm walking in a faerie tale.*

---

* Humans have been scrutinizing dreams for millennia for secrets that might benefit us; there is evidence that ancient Sumerians and Mesopotamians were versed in dream interpretation. Modern dream research on the neuropsychology of dreaming affirms that dreams relate figuratively to actual events in an individual's life, and their nonlinear appearance and unfolding are a source of creativity and innovation in waking hours.

It is a lovely piece of knitting, though I say it myself. People who see me in the grocery store or in the library or out on an urban walk will often say, "I love your..." then trail off, seeking the mot juste. Because, let's face it, we don't wear a lot of hoods these days, and this one hangs far down my back with its fringed tassel and bell, a mysterious thing. "Hood?" I offer, and the suggestion is taken up gratefully, followed often by a discussion of knitting patterns, woolen yarns, or per-haps dragons. (For as Ursula Le Guin wrote, "People who deny the existence of dragons are often eaten by dragons. From within.")

On a cold day at the wooded park near my home, I had the hood pulled up over my ears and almost down to my sniffly nose. I stopped at a twisted old cherry, covered with the neat lin-ear sap-well holes of the red-breasted sapsuckers who have been feeding from this ancient tree for decades. I placed my hands on the trunk while peering around the other side, where I was startled to see a sapsucker working the bark. In spite of the bird's name, red-breasted sapsuckers in the Pacific North-west have rather yellow breasts but entirely red-feathered hoods, covering their heads, chins, and napes. Why had this bird not flown off at my approach? We startled each other, but neither of us left. Soon she went back to her work, in spite of my laughter. Two red-hooded beings, loose in the urban wil-derness.

The Wildflower Task

Jungian psychologist and folklorist Clarissa Pinkola Estés teaches that the wolf in the story of Little Red is The Predator—that thing which lures all of us, at some point, away from our highest path and truest self. From this perspective, we learn to avoid this wolf as part of our maturation. I love Estés, but my reading of this tale is different.

When the wolf lures Red off the trail, he sets her a task—something to occupy her time so that he can pad over to the grandmother's house and eat her up. The task is picking wildflowers. Now, instead of just a boring tin of muffins for her grandmother, Red will show up with a basket full of wild beauty that she gathered herself from the forest—treasures not found on the well-traveled path. She will find her own way through the woods. And when she arrives? Yes, she will be eaten by the wolf, as her grandmother was before her. But it turns out that they are both perfectly well. They are swallowed whole by the wild and emerge *exhilarated*. They place the flowers on the table and feast with an exquisite new hunger.

I am positive that at the end of the tale, when Red promises her wise elder grandmother that from now on she will "stay on the path," it is with a wink and a nudge between them. The grandmother has known for some time, and Red is just learning: Who wants an everyday path—paved and void of danger—when we can have beasts and shadows and secret flowers and unexpected visits from the feral wolf of our imaginations?*

---

* As far as historical accuracy goes, Estés's reading of the wolf as predator is correct. The earliest versions of this tale are violent: more than the varnished

The "I Am Here" of Hope

The ritual of entrance into most monastic orders is kept largely secret from the public, but a Benedictine monk friend told me about one of the most powerful moments in his own profession of vows. Benedictines pledge to an outwardly paradoxical life of simultaneous conversion and stability—a constant evolution of mind and spirit within the rootedness of a particular community. When asked a series of questions about the uncompromising spiritual commitment he was vowing to make, my friend's ancient traditional response was one that offered no space for irony or hesitation. The word is Latin, from monastic ritual dating back to the sixteenth century: *adsum*. I am here.

Alongside all the bright studies that affirm our connectedness to nature and its health benefits for human mind and body, there are new studies about the health of the earth as a whole. Ecological crisis rushes upon us faster than even climate scientists were expecting. I don't need to rehearse the roster of wreckage and urgency—it is well documented and updated constantly. By the time this book is out, the fears that rise most swiftly to mind will doubtlessly have grown and changed. As I write, the Amazon burns, the western states where I live are ablaze with wildfires, and the Salish Sea, my home waters, are polluted by the effluence of six million humans in the greater Seattle area that surrounds it, some of this sullage my

---

story we find in today's children's books of a grandmother and a girl who spring unscathed from a wolf's belly, there is the intimation of the girl's sexual assault. My reimagining here is in part an effort to actively disconnect the role of the wild wolf as an essential ecological presence from the role of sexual predator.

own. Sea stars are literally melting with wasting disease, and our beloved southern resident orcas are dwindling and thin. Responses to all of these things are forestalled by the coronavirus pandemic, which instills fear while emphasizing our interconnected vulnerability. Everyone reading this has their own current litany of global and local earthen anxieties. Our hearts are breaking.

Yet the profound political, cultural, and personal will to make the radical changes that will result in a prevention of ecological collapse continues to show a lack of sustained depth and commitment. Solutions to the current crises appear beyond rational hope. Somehow, in all of this complexity and loss and seeming hopelessness, a response is asked of us, and in this response lies the seed of our own rootedness.

When Rachel Carson's beloved friend Dorothy Freeman inquired in a letter why Rachel must delve into the unpleasantness of the "poisons book" (as they referred to the yet-untitled *Silent Spring*), Rachel responded that she felt called by the dead birds she'd held in her hands, and could not live, knowing what she had learned about DDT, without speaking, without — her gift — lifting her pen to write. *Adsum.**

---

* There is a feminist critique of the tendency among writers, no matter what gender, to refer to women writers sometimes by their first names, but men always by their last names; this is interpreted as an act of diminution somehow. For me, it is the opposite: an expression of love, knowledge, intimacy, and inspiration. Rachel Carson's books and letters have been my constant example for decades. Kristin LeMay writes in her remarkable book on Emily Dickinson, *I Told My Soul to Sing,* "Emily has become for me a sort of unlikely patron saint, and we don't call the saints by their last names." So, too, with Rachel for me.

The *adsum* of monastic profession is not the "be here now" of pop-mindfulness practice. In this context it indicates a radical openness to an entire life of psychological wilderness, one of few comforts and constant uncertainty. The life, in a form, we all live. *Adsum* abandons hopelessness and blind hope and even rational hope—which, of course, are not the only kinds. I discovered this visionary definition in a Benedictine monastery library some years ago: Hope is "that virtue by which we take responsibility for the future," and a quality that gives our actions "special urgency."

While "virtue" might sound a bit prim and moralistic, I looked it up in the same old monastic lexicon and found virtue described as the power to realize good, to do so "joyfully and with perseverance in spite of obstacles." Hope asks something of us. The singular virtue of acting in hope has nothing to do with the likelihood of a specific outcome; it has simply to do, in this moment, with participation in the renewal of the earth, however that will manifest.

Rebecca Solnit writes in *Hope in the Dark,* "It's important to say what hope is not: it is not the belief that everything was, is, or will be fine." Hope is not a remedy or even a substitute for the despair and anxiety we face in the modern world, but a companion to these things. Mature hope involves a willingness to allow that brokenness and beauty sometimes intertwine. We live in "an extraordinary time full of vital, transformative movements that could not be foreseen," argues Solnit. "It's also a nightmarish time. Full engagement requires the ability to perceive both." She speaks directly to the pairing of complex hope and the summons to respond:

The hope I'm interested in is about broad perspectives with specific possibilities, ones that invite or demand that we act. It's also not a sunny everything-is-getting-better narrative, though it may be a counter to the everything-is-getting-worse narrative. You could call it an account of complexities and uncertainties, with openings.

Julian of Norwich, anchoress, writer, and mystic of the English Middle Ages, proclaims in one of her most famous "showings," or visions, that within the sacred whole, "all shall be well, and all shall be well, and all manner of thing shall be well."* So often this is read as a template for infinite consolation, as if the saint is patting us on the head and telling us in a calming nursery school marm's voice that it's "all going to be OK." But this is not so. Julian's "wellness" neither transcends human error nor absolves it, but awakens us to the truth that we are called to loving action within the complex awareness that all things are blessed in a wider fullness, whether events unfold according to our hoped-for outcomes or not. Julian herself never lingers in certainty. After a particularly comforting vision, she writes,

> My soul rested in a state of such delightful security and powerful bliss that there was no fear, no sorrow, or physical pain that could be suffered that would have bothered

---

* An exact contemporary of Chaucer, Julian wrote in Middle English, which requires translation for most modern readers. This line is one of the few that is easily readable on its own.

me. And then again I was suffering. And then I was in bliss.
Back and forth — first one, then the other.

She concludes that in the sacred whole we are equally supported "in well and woe . . . both are one love." Uncertainty feeds
our intellect and our actions; without it, we would have only the
simple banality of "knowledge," with no need for the unsettling
motivation of hope at all.

Pilgrim on the Wolf Path

Who among us has not heard it? The wolf of this beloved, damaged earth, beckoning us by name just outside our safe living
room, demanding our own response? The strange and persistent
furry-pawed knocking? We peek tentatively through the door,
just ajar, and see that there is no road, no sidewalk, barely a
trail — and that obscured by stones, by leaves, by an intimation
of the remains of those who have walked before us upon the
unyielding circle of life.

In spite of it all, we long to walk this path. For we know that
there is more than what has been given and named by the over-
culture, more than what we have been told is true, more than
green gardens and nature calendars, and recycling, and a summer hike in the mountains, and an occasional camping trip.
More, even, than an hourlong "forest bath," however lovely that
sounds. We know there is a wilder earth, and upon it — within
it — a wilder, more authentic human self. *We know the need of
each for the other is absolute.*

Adsum. We pack our satchel lightly and cross the threshold.

We enter into a geography of mind and soil, unsteady as we track a wilder foothold.

We enter as pilgrims, as wayfarers — knowing there is something we are seeking, something nameless, beautiful, waiting, wanting. Something that will change us so thoroughly that our cozy slippers will no longer fit, that our cat will, at first, hiss upon our return, our hair tinted green with lichen, sweet root tendrils among our toes.

We enter knowing the path is invisible, the terrain uneasy, the weather uncertain, our minds unprepared.

We enter knowing we are not able to "save the earth" in one fell swoop or any swoop at all, that our work here will fall short even as there is no alternative but to act.

We enter absolutely entangled with every creature, tree, stone, and human above and below the soil. And with this as our lodestar and comfort, we understand that we enter, too, completely alone, with little but our wild instincts to guide us.

Some ecological activists want to reserve the word *wildness* for the material, nonhuman, big wild — the forested wilderness, the desert, the ocean, and all the planet's creatures whom they work so hard in their lives to protect. There is beauty in this impulse, but I believe that it is also important to claim wildness as a state of mind and a way of being. Both dimensions of the word — the wild as sacred lands and the wild as quality of consciousness — must inform the other: wild hearts breed love and protection of a wild earth.

Listening to the Wolf

This sensibility grounds another famous wolf story from the western European canon. The thirteenth-century Italian saint Francis of Assisi is the patron of ecology, beloved far beyond the Christian faith. Francis sings praise to the divine, in his words, "through our Sister, Mother Earth, who sustains and governs us." Note that Francis does not speak of an earth that is a gift to humans from a creator-God, nor a garden of resources over which we have dominion, nor even a landscape that we are called to steward benevolently. It is the earth, a mother, our sister, who *governs us.* We find the sacred not simply upon the earth, but *through* the earth. Francis was known for calling all things sister and brother—not just the monks and sisters of the monasteries he cofounded with Clare of Assisi, but everything. The sun, the moon, famously. But also crickets, grasses, mice, trout, ravens. Wolves.*

Francis was called upon by the mayor of Gubbio, a prosperous hillside town. Upon Francis's arrival, the mayor begged him to intervene in an unfolding tragedy: a ferocious wolf was padding the village edges, slaughtering livestock, attacking shepherds, and killing the guards sent to slay her. The more the stories of savagery passed along the villagers' tongues, the more enormous, fierce, and bloodthirsty the wolf became. Everyone lived in constant dread.

---

* Francis is said to have allowed a library full of sacred texts to burn to the ground to keep from harming "brother fire" with an onslaught of water. No one is called a saint for being normal.

Commentators who gloss the fable consistently use the verb *tame* to describe Francis's interaction with the wolf, but my research into longer tellings of the story reveals that Francis did not solve the conflict by taming the wolf at all. He solved it by *listening* to the wolf. He stood in the wolf's presence, respected the animal's wildness, apprehended her story, called her Sister.

Francis returned to the village with news that the wolf had been injured, and abandoned by her pack. She had not meant to anger the people, she was simply hungry, and sheep are so slow. When the guards spotted the wolf and ran at her with their spears, she had acted in self-defense. Armed with Francis's understanding, the townspeople helped the injured wolf find food, and they coexisted without fear.

A Wild Summons

Rachel Carson and Dorothy Freeman both loved *The Wind in the Willows,* by Kenneth Grahame; in their decadelong correspondence (ending with Rachel's death in 1964) the friends shared thoughts about the book frequently, and they read aloud from it when together. I was delighted to discover that Rachel's favorite chapter is the same as my own: "The Piper at the Gates of Dawn." In one of the most wonderfully mystical passages in all of nature literature, Rat is out all night in his little boat with Mole, searching for Portly, the baby otter, who is lost. Up and down the wooded riverbank they paddle without success until finally, at the liminal, suspended time of night-into-dawn, they hear floating from the woods—and just beyond their rational apprehension—a fluted song.

The notes drift from the pipes of Pan, mischievous faun-guardian of the wild forest. "O Mole," Rat cries. "The beauty of it! The merry bubble and joy, the thin, clear, happy call of the distant piping! Such music I never dreamed of, and the call in it is stronger than the music is sweet! Row on, Mole, row! For the music and the call must be for us." Portly, of course, is safe with the faun. Meanwhile, the mystical oneness that has overcome the friends grows foggy and dim as the sun rises and every-day life intrudes. But I agree with Rat: *The song and the call is for us.*

Rat lives in a riverbank hollow. Mole lives in a burrow below-ground, just north of the river and the Wild Wood. Rachel Carson lived in a suburb of Silver Spring, Maryland, and summered in her Southport, Connecticut, cottage. I live in a Seattle neigh-borhood, an urban place interspersed with woodlands that skirt the Salish Sea, and just a short ride from the various buildings that house Amazon headquarters. The editor of this book lives on the fifteenth floor of a Manhattan apartment building. Yet Rat speaks the truth: the call is for us. All of us.

The nature of urban places simultaneously enchants us, edu-cates us, and points us to the ever-presence of a wilderness beyond what we will ever see or experience. With attentiveness we discover that the places humans inhabit are far more biologi-cally diverse than we imagined. With all the earth we share the presence of moon, clouds, weather, rooted beings, branching beings, winged beings, antennaed beings. Our windows open to wind; potted plants bring a tangle of roots to our slender windowsills; hummingbirds visit our fuchsias; raccoons pad our early-morning sidewalks while we sleep; there is a chaos of life beneath our feet—even when the soil is covered with concrete. We attend all of these things with wonder.

Yet we must be careful not to allow the delight we discover in the urban wilds to lull us into complacency. One of the most significant lessons of urban nature is the invitation to think beyond the concrete. When we cross the threshold where the sidewalk ends and witness the sudden leap in the diversity of plant life and birds (other animals, too, more hidden), we are reminded that while the urban wild is meaningful, it is not, and can never be, *enough*.

New work in psychology reveals that simply the *knowledge* of a flourishing wild—beyond the reach of most human visitors— positively affects our psychology and creativity. Wilderness *that we will never even see* roots us in a sense of individual purpose, no matter where we live. It is this knowledge that we are called to cultivate daily. We live in a continuum. From the micro-ecosystems of our households we make the choices that will allow the endurance of deepest wilderness.

There is a reason this chapter begins with a faerie story. "The forest is the place of trial... dangerous and exciting," Sara Maitland writes in *From the Forest: A Search for the Hidden Roots of Our Fairy Tales*. "Coming to terms with the forest, surviving its terrors, utilizing its gifts, and gaining its help" is the only way to the always complicated "happily ever after" we find in such myths.

While the wolf in these pages appears to be metaphoric, she is a mirror for the furred wolves of blood and bone who, clawing for their place and their continuation, track the diminishing land. The untamed wolves of earth and imagination stand together with a summons, howling for their need of our lives. Step outside, or lean your head through an open window. Affirm the earth beneath you, let the sky touch your body. Now breathe, and listen. Just listen. It's okay, for now, if only a wisp of a voice reaches your ears. Answer out loud anyway: *I am here*. Our first step acknowledges the call to a wilder path — the offering of our own *Adsum*.

Our second step is to begin walking it. And here we leave metaphor completely — for the beginning of walking a path is, in fact, walking a path.

LISTEN FOR THE WILD SUMMONS

Shed

Bare Feet, Holy Ground

One sunny spring day some years ago I was walking bare-foot in the wooded park near my home. For some reason, the Seattle Parks officials see fit to maintain the central trails by layering them with sharp-edged stones that become wedged and stabby on the path; I've discovered a side trail that gives me a good half hour of barefoot walking upon soft soil made up of fragrant decomposing redcedar, fir needles, and fallen leaves.

On this day, I passed another woman with bare toes who had discovered the same trail, her running shoes tied together and hanging from one finger by their laces, just as mine were. She glanced at me, a beatific smile upon her lips. "Happy barefoot walking," I said. I suppose silence would have been more in keeping with the spirit of the moment, but that's what popped out. She walked serenely past while saying, "I'm earthing."

Earthing. Well, that sounded lovely—I searched the reaches of my little brain and found the word, a trend from some time ago. But in the moment, I didn't need to know the origin of the word or the science behind the fad (if there was any science—I wondered). I knew exactly what she meant. She was doing more than going barefoot for fun, or in celebration of a warm day. She was consciously connecting with the earth through her footsteps. "Well, then I'm earthing too," I said quietly to myself. And as so often is the case, speaking it made it so.

A book on the subject of earthing first appeared in the 1980s, and was authored by an unusual trio: Clinton Ober, a cable TV salesman; Martin Zucker, a writer on the subject of natural healing, fitness, and alternative medicine; and Dr. Stephen T. Sinatra, a cardiologist and bioenergetic psychotherapist. Together they published *Earthing: The Most Important Health Discovery Ever?* A year later, the trio had gained in confidence and were emboldened by publication of papers in the *Journal of Environmental and Public Health* (a peer-reviewed journal, but not a scientific one). The next printing featured the same title with a tiny, jubilant change: *Earthing: The Most Important Health Discovery Ever!*—the question mark replaced by an exclamation point.

In a nutshell, the book suggests that walking barefoot on natural surfaces, even just twenty minutes a day, balances our bodies' positive ionic charge with the earth's predominantly negative ionic charge, helping to prevent inflammation and its attendant ailments. (For those unable to walk, it is suggested that placing feet on the earth without walking works just as well.) Earthing is said to reduce pain in the feet and other parts of the body, prevent or improve a host of diseases and neuropsychological issues, accelerate wound healing, calm anxiety, and promote better sleep. I scoured all the papers I could find, and spoke with physicists, geologists, and electricians. As much as the barefoot wanderer in me wants this to be right, it's my feeling that the electromagnetic aspects of the science remain murky to date, with self-referential evidence, and subjective research seeking support for a presupposed conclusion, none of it congealing quite into a convincing whole. Maybe it will one day.

Meanwhile, there is another scientific avenue that does support the benefits of barefoot walking upon soil, leaves, grasses, exposed roots, and sand. Biomechanists and exercise scientists demonstrate that walking shoeless on varied outdoor terrain strengthens muscles, corrects bunions, and increases grace and ease in movement of all kinds—a physical intelligence accessed from the ground up.

What We Leave Behind

In the Hebrew book of Exodus, when Moses approaches the burning bush to receive the commandments, the divine Voice orders, "Shed your sandals." This is ordinarily translated as "take off" or "remove" your sandals, the way we would typically talk about our own shoes. But the original verb is more radical. *Shed*. Cast off. Leave behind. This is the verb we use for the great transition of animals, the shedding of fur, the sloughing of an entire snakeskin, the emergence as the same creature, made new.

And that which is shed? Like the antlers of a deer, they are the things that were once an organ of our lives but are no longer of use. Otherness. Separateness. Elevation. Pretense. A certain kind of beauty. A certain kind of belonging. The certainty of comfort.

Ecumenical Trappist monk Brother David Steindl-Rast notes that the passage in Exodus continues: "Shed your shoes; *this is holy ground*." We tend to think this means that Moses is to remove his shoes *because* the ground is holy, a matter of respect or humility, perhaps, as one is asked to remove shoes when entering certain mosques or temples. But according to Brother David, the rabbinic scholars he has spoken with insist that it is the other way around: "When you take off your shoes, you will *notice* this is holy ground!... because what prevents you from seeing that it's holy ground is the dead skin you have to shed." It's only our own limited thinking (and lack of proper shedding) that keeps us from the apprehension: *we are always standing upon holy ground*.

I feel this very literally walking shoeless. Barefoot walking increases my ecological knowledge of the places that I walk, in unexpected ways. While watching where I step, I find feathers, fungus, tracks from the tiny feet of my animal-cousins, and all manner of small living wonders. Caterpillars in their instars, always shedding. The rough-skinned newt on his overland migration. A hummingbird eggshell, broken open and impossibly small, and then, searching, the nest above it with tiny pin bills pointed skyward.

Shoes put little blindfolds on our soles. Shoes shed, I feel the earth change underneath my feet—the places where the soil turns suddenly extra cold, say, and I look up to find a closed cedar canopy; I realize I am standing on soil that is *never* touched by the sun, and looking down again I see what happens here—

the hefty growth of sword ferns, the lack of nearly every other plant.

And if I am very quiet and still, and let my feet stand solid upon the earth, I can feel what I have come to call *beneathness*. For the soil is alive and writhing beyond my sight with roots, mycelia, decomposers, bacteria, protozoa, worms, grubs, beetles— beyond counting, beyond knowing. The living and the dead brushing together to create the quietest symphony of sound and activity. Holy ground.

The Megalithic Measure

During the writing of this book, I traveled to Ireland, and I am still hushed and wondering over the mystery of the megalithic sites—the passage graves at Newgrange and Knowth of Brú na Bóinne, the stone dolmens that stand as sage presences upon the landscape, alongside the sheep. Entering the Newgrange passage, I was visited by an awe I have rarely known in the whole of my life, and I feel the tendrils of this place's mystery—so of-the-earth yet so beyond my understanding—still unfurling in my body and mind. I struggle to comprehend the five-thousand-year span that has elapsed since the creation of these enigmatic structures, my own wandering there, and the lessons it all holds for my small life.

The stone formations are situated to interact with aspects of the seasons and astronomical events. At Newgrange, most famously, on the winter solstice the inner chamber is illumined with a shaft of light that beams all the way from a tiny rectangular opening at the passage entrance to the interior graves. Celtic art scholar Adam Tetlow writes that the passage graves were constructed using a plan based on what we would now term whole numbers, grounded in a pacing of footsteps (what became in English measure the *foot*), thus "linking the proportions of the body to the dimensions of the earth." It is likely that this ancient culture did not know numbers, and so could not "count" their paces. It is difficult for us now to imagine measurement without numbering. Somehow footsteps offered an innate compass, connecting humans to both the earth and the cosmos.

I removed my shoes at Newgrange, placed my bare feet on the mingled stones and grasses, and felt suddenly nauseous — the direct contact with ancient soil called to dizzying consciousness the countless footsteps of women like myself who came before — working, birthing, laughing, mourning. And the creatures, too, for a ranging megafauna once roamed a now dewilded Ireland, including giant stags, and huge endemic bears whose footprints were shaped much like our own.

Walking Clay

For millennia, and across cultures, bare feet were a sign among monastics of humility. The words *humility* and *human* are both rooted in the Latin *homo,* meaning the ground or the earth — the *humus.* The late Irish poet and philosopher John O'Donohue called upon this connection in his Celtic spirituality: "Fashioned from the earth, we are souls in clay form."

Buddhist and other monastic traditions often shed shoes in favor of barefootedness or simple sandals, and for someone who was observed levitating above the earth on several occasions, the great mystic Teresa of Ávila was uncommonly attentive to the state of her spiritual sisters' feet. When she and her *anam cara,* the Spanish wildman-poet John of the Cross, reformed the Carmelite monastic order in the 1500s, they took it as their first priority to get rid of the fancy Spanish shoes that were the norm in rich convents. They formed a simpler new Order of *Discalced*—unshod—Carmelites.

Monastic humility is not a form of meekness, or false modesty, or subservience to higher authority. The humusy, earthen humility sought here is one of sweetness and of foundational strength.* O'Donohue writes: "Your feet bring your private clay in touch with the ancient, mother clay from which you first emerged." This is the ground of remembrance—we belong to this earth that cradles our beginning and our end, clay upon clay. Barefootedness can be an honoring of one's own closeness to the cycles of nature, the inevitable interweaving of soil and self. *For dust thou art, and unto dust shalt thou return.*

Thomas Merton enjoyed regular barefoot walks in the woodlands of his Trappist monastery in Kentucky, his soles upon the soft pine needles invoking a sense of the sacred—the "presence

---

* The word *sweetness* has itself come to be conflated with a Hallmark-saccharine sort of sentimentality, but I want to reclaim the earlier, spiritual meaning—*dulcis*—a sense of sacred presence, and the uncommon peace this brings.

of the Presence." Buddhist monk Thich Nhat Hanh exhorts us to "walk as if we are kissing the earth with our feet." Such walking is a service, a blessing, an offering, a prayer.*

Intelligent Feet

Shoes have obvious benefits for protection, injury prevention, and, of course, the artistic self-expression found in fashion. I have friends who would rather cut off a hand than give up their Jimmy Choos. But in other ways shoes have tamed and hobbled us. I am obsessed with the work of biomechanist Katy Bowman, who says her research on the movement of the human body brought her to the study of ecology.

Bowman finds that in the modern "walking for fitness" mindset we measure the intensity of our movement in terms of speed and number of steps—all of this is typically assessed through the window of people walking (or running) in super-protective, ultra-engineered shoes on an even concrete substrate. In her book *Move Your DNA*, she observes that when we calculate the number of steps per minute required for peak fitness benefits, we do it by determining the optimal rate for walking upon "bland, unchallenging, and mind-unnecessary environments that cheat

---

* Brothers Nhat Hanh and Merton met once, in 1966, at Merton's monastery, the Abbey of Gethsemani, where they broke bread together in a private, ecumenical ceremony of communion. Nhat Hanh, who had taken a vow to drink no alcohol, convinced Merton to offer grape juice in the ritual, instead of the traditional wine.

us of many of the benefits of walking." *Mind-unnecessary environments*.

Bowman finds that when she invites her students to remove their shoes and walk on the uneven surface of a forest floor, they walk more slowly. But does this mean they walk with less work? It does not. Speed and steps are not the only measure of fitness. There is also the challenge of varied terrain, fluctuation of grade, sensory involvement with our steps, awareness of landscape, relationship of substrate to overall body-mind agility. All of these add to both physical fitness and mental acuity. When I interviewed Katy, she told me, "The ability for bare feet to successfully negotiate the earth is a humanity-old skill. It's a natural intelligence."

While living in the Sangre de Cristo mountains of New Mexico, Clarissa Pinkola Estés learned of the Southwest archetype La Que Sabe, The One Who Knows. La Que Sabe created humans from "a wrinkle on the sole of her divine foot." We are knowing creatures, in part because we are made "of the skin of the sole, which feels everything."

There is an innate connection between the soles of our feet and the soil that brings recognition, enlivened spirits, and unexpected joy. Our bodies and minds are made from, and *for,* this discerning intimacy. Yet most of us in modernized cultures have forfeited this indwelling intelligence.

Human Paws and the Dishonorable Toe

We are the only species that imprisons our feet in shoes—and not always because we need to. On most substrates, and in temperate climates, our walking directly upon the earth with bare feet or minimal footwear gives rise to a gait of ease and grace. We develop calluses, giving our soles more the feel of a raccoon's pads than the baby-bottom softness we insist upon in overprotecting and ultramoisturizing our animal feet.

My own feet are small and pretty (if I may say so) in spite of their pudgy toes that look like sausages. My baby toenail is so tiny that when I polish my nails I actually have to paint a little piece of my toe to get some color there. But the bottom of my feet? They look like they belong to a cloven-hoofed beast.

Some people break their toes honorably, while climbing stone faces in the wilderness or saving children about to be hit by a car. I broke mine after sitting cross-legged on the living room rug, folding laundry. There was a ruckus in the kitchen, which I believed to be our tuxedo cat, Delilah, attacking and certainly gruesomely killing Carmen, the starling who lives in our house. Usually I keep cat and bird separate, but I must have left the door to Carmen's aviary open while Delilah was prowling for dinner. I jumped up, a wild hare, lurching to the kitchen to rescue Carmen and slamming my toe along the way onto the unforgiving corner of a hard wooden chair leg. Swallowing the pain, I limped to the aviary and found Carmen serenely preening, door closed.

The sound I'd heard was probably just her flapping among her toys.*

I looked down at my foot, already swollen and an odd shade of lavender. It would run through every color of the rainbow in the next weeks. When I visited my doctor with the hope that my big toe was just bruised, she pronounced a likely stress fracture. She toyed obliviously with my aching purple toe while musing on the typical treatment: ice and extended rest. As adhesions grew around the small fissure it would heal, but I would likely feel some pain after extended use for, well, forever. She suggested rigid-sole shoes to limit movement in the joint.

"Great," I thought. I had a cute pair of wood-soled Danskos in the closet, that I'd forgotten about. This would be the perfect time to drag them out.

But after spending so many years taking barefoot walks and wearing shoes with minimal footbeds, the inflexible soles were completely disorienting. Wearing the clogs on a forest path, I might as well have been in the Target parking lot or on an ice floe in Greenland. I felt unbalanced and cranky.

I stuck with them for a couple weeks to see what would happen. While my foot did start to hurt a little less, I soon discovered that the mobility of my toe was almost entirely lost. Dense adhesions were growing all around it as my doctor had predicted; the podiatrists I spoke with told me these were likely permanent, and a positive development. Protective.

---

* Carmen joined our household while I was writing the book *Mozart's Starling*. More to come on Carmen in these pages.

But as human animals, the mobility of our hands and feet gives us one of our greatest ways of apprehending the world—a means of movement, play, and sensory detection that feeds our creative ingenuity. When I asked Katy about all of this, she told me that "casting" feet in shoes that restrict movement is beneficial in acute situations (like a freshly broken toe), but after that the model breaks down. She works with podiatrists to help patients do the slow work of returning full, natural motion. "The idea that stiff containers are the safest environment for feet runs quite deep," Bowman told me, "but today's shoes have moved beyond protection to a place where we've induced a way of walking that prevents our feet from moving much at all."

As the fracture became less painful, I replaced the clogs with light runners and added the slow, progressive exercises recommended by Bowman to my toe-healing regimen.

I returned to my earthing path. It hurt a little, but I walked slow. Super slow—up to fifteen seconds on one injured-footstep, letting each tendon spread and lengthen, letting the uneven earth, smooth stones, and twisty roots allow more natural movement and mobility to take hold. I started with five minutes and a few yards of such walking and worked up in the course of months. Now I regularly walk a couple of miles or more with no shoes on natural terrain. I look down at my wild, ranging foot, doing what feet do best—moving with joy and gratitude and blessing upon an ever-changing path.

Homecoming

Christopher McDougall's excellent book *Born to Run* had an unintended side effect: the spawning of a barefoot running movement that led to thousands of injured runners limping into doctors' offices with stress fractures. This in turn gave barefootedness itself a bad name. Of course, after years of wearing shoes, suddenly running barefoot on paved sidewalks is a punishment to the intricacy of a foot, which is strong but sensitive—evolved to be both things at once. In her biomechanist's lingo, Bowman encourages "small increases in loading behaviors" that, over time, will allow our feet to undo the constraints of traditional footwear without injury.

Barefoot walking is a wonder and a homecoming. "For the same reasons folks seek to protect heritage seeds or their culture's art forms," Bowman told me, "walking barefoot is how we preserve the footsteps of our ancestors at the same time we preserve our own." We start slow, shedding our shoes for a few minutes each day, feeling our feet on soft carpet, then soft grass. Eventually a morning tour of the backyard or nearby park or grassy apartment building parking strip without shoes, taking months or longer to work up to a half hour of walking on a natural surface. Of course, not everyone has access to a backyard, or even grass. Thomas Merton wrote in his diary that the benefits of shedding his shoes could be experienced even indoors. On Ash Wednesday the monks would approach the altar unshod: "Going barefoot is a joyous thing. It is good to feel the floor of the earth under your feet"—a lovely reminder that the floor itself is made of the elemental things, just as our bodies are. We are not plum blossoms. Just because it is a bit chilly or damp is no reason to keep our shoes on. Cold is enlivening and rarely embraced, awakening our nerve endings and bringing awareness of the day—always changing throughout hours, throughout seasons. When protection is sensible, our feet can be at home in soft-soled moccasins or other simple shoes that let more of the world in to inform beautiful, full movement. Changes of terrain, temperature, grade, and substrate engage the ecology of body and imagination.

We mind where we step, deepening our kinship with all the beings who share our earth and are responding to these same variables—those with roots and paws and delicate bird feet. All standing upon holy ground. All shoeless, like us.

GO BAREFOOT

Wander

A Feral Cartography

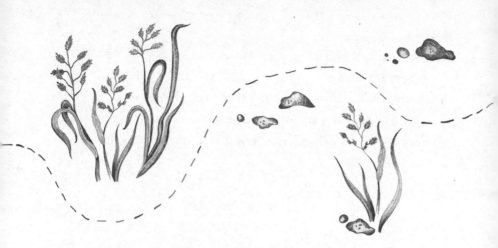

W*ander* is a beautiful word—a word that startles our imagination into wanting. And longing, too, for in a world of schedules and deadlines and Google Maps asserting a linear, time-bound path through our lives, the idea of wandering seems a sweet dream, ever out of reach. Wander*lust,* ever unrequited. Who has the means? Money is not required, nor equipment of any kind. The necessary means here are more rare: A spaciousness of mind. An expansiveness of time. An unhurried pace. It's a circular problem to be sure. While these things are essential for good wandering, it is wandering itself that engenders them. Sort of like the conundrum of morning: How are we supposed to make coffee when we haven't *had* coffee?

Walking is the naturalist's pace—the tempo at which observation can occur. The sniffing, watching, inhaling, uphill, downhill, spiral pace of footed mammals. Alongside certain ways of running and dancing to rhythm, it is among the ancestral, wild movements for humans. Walking, we are self-reliant, knowing inhabitants of our bodies, and in cahoots with other animal bodies. But where walking is a way of moving oneself through the world, wandering is more of a mind-set. In *Wanderlust,* Rebecca Solnit writes of her suspicion that "the mind, like the feet, works at about three miles an hour. If this is so, then modern life is moving faster than the speed of thought, or thoughtfulness." Wandering brings mind and movement into a healing congruity.

Skeletons

The root of the word *wander* lies in the Old English *wandrian* — to wend, to wind. Wandering is goalless, aimless, directionless — the motion of our steps guided not by a set path or route, but by the inclination of our spirit. Off grid, offline, off map. We may set forth to wander with a light pack or with nothing at all.

This was the spirit of Bashō, the seventeenth-century Japanese poet-recluse, who set out on an indefinitely long ramble, relying upon hospitality from monks and fellow poets along the way. He'd read Lao-tzu: "A good traveler has no fixed plans and is not intent on arriving." In fluid brushstrokes, Bashō professed:

Following the example of an ancient priest who is said to have traveled thousands of miles caring naught for his provisions and attaining the state of sheer ecstasy under the pure beams of the moon, I left my broken house on the River Sumida...among the wails of the autumn wind.

Switching to verse, Bashō became

Determined to fall
A weather exposed skeleton.

Bashō had been depressive in the face of town-living and apathetic about his fame among the literary cognoscenti. After the death of his mother his acedia worsened, and his disciple-students proudly built him a new hut in Fukugawa, intending to ground and cheer the poet. Here Bashō refocused on Zen meditation and planted banana (*bashō*) trees, but nothing seemed to calm his spirit or deafen the self-scrutinizing voices in his head.* In a time when few wandered for simple health or pleasure, Bashō and his small reed-woven satchel set out on a spartan and dangerous journey from Edo (present-day Tokyo) toward Mount Fuji and on to Kyoto, fully expecting (and, in his melancholia, likely half hoping) to die of exposure, or to be murdered by bandits.

But neither of these things happened. As he rambled, the darkness of spirits that had plagued him for years lifted; brighter and brighter his heart became as the journey lengthened and deepened. His poetry, too, became even more penetrating. Of course, I do not wish any of us to fall as skeletons on our wanders—at least not today. But Bashō's abandon is both edifying and transformative.

Wandering tilts us out of the everyday measure of chronological time and into the eternal spiral of *kairos*—sacred time. Our

---

* Bashō the name and *bashō* the plant are homonyms, but the Japanese characters, or *kanji,* used to write the names of poet and plant are different. This wordplay is said to have delighted the poet.

typical walking to a particular destination can hinder us from abandoning ourselves to the moment. In wandering, we give this up. Footsteps are decommodified: we remove the Fitbit as we would take off our shoes to enter a sacred river, for that is what we are doing—stepping into the place where the wild earth meets our wild minds. Holy ground.

Dangerous Business

"It's a dangerous business, Frodo, going out your door. You step onto the road, and if you don't keep your feet, there's no knowing where you might be swept off to." Wandering, we are feral—we have escaped. Anything can happen. Ominous things. Luminous things.

The unpredictability of a wander brings depth to the predictable elements of life. When there is no anxiety to reach a particular destination, there is an opportunity for adventure. Explorer Alastair Humphreys's notion of microadventures is at work here—adventure need not mean a Tolkienesque quest or rigorous physical danger (though occasionally such things are good for us). It can be just stepping through the portal of our doorways with a spirit of openness.

Even close to home, the wandering way is entirely undomesticated. What might it bring? I list here only a few of the small adventures I know from experience: an enlivening flirtation; a fritillary visitation; a desiccating-worm rescue; the discovery that a neighbor who strutted out of his truck one day with a brace of extremely dead wild ducks of various species also works tirelessly to outlaw leg traps and gardens an expanse of lavender. A conversation with a sparrow in language suspended sweetly between the human and the avian. A mind unclouded enough to answer when a tree beckoned me by name, when grasses demanded a shedding of shoes, when never-known thoughts arose unbidden. An unplanned nap in a field—lying down without meaning to, and so greeting warblers overhead, just arrived on their migration from Central America and glowing tropical yellow.

Beware! One step out the door and it is easy to fall into established routes. I have mine: down the sidewalk stairs, through the small cedar copse, across the meadow, past the owl tree, back through the forested earthing path....It takes pausing and remembrance to wander.

Last week I set out in the early morning and (in spite of being in the midst of writing a chapter about the benefits of wandering) absently fell into my ordinary course, when I spotted a small dog with big ears. No, I spotted a large cat with lopey feet. Why does it always take a long second, no matter how many times I see one on the neighborhood sidewalk? I spotted a coyote. I see this little girl now and then when I get up early enough, and always she disappears around the same corner. Knowing her usual route, I decided to figure out where she was going (from a healthy distance—we don't want urban coyotes to normalize human proximity). It was not easy. Though outwardly appearing to amble at a comfortable pace, she is uncommonly speedy. And though I got farther than I ever had in coming to know her path this day, still I lost her fairly soon.

Coyotes are expert at magically disappearing—slipping into

the grasses, the thin spaces between trees. But on a sidewalk, where could she have gone? Well, lots of places, actually, with all the weed-grown alleyways. I looked around, pondered the most likely pathway, and followed it. At every corner, and at the end of each little backstreet, I'd look up and think, *Where from here?* I did not find the tiniest hint of being on the coyote's trail. But at one point I lifted my head to ask again, *Where from here?* and realized that I had no idea at all where *here* was. I have lived in this neighborhood for sixteen years, and in just twenty minutes I found myself at a crossroads I'd never been at before.

Giving up on my ill-conceived coyote tracking, I continued to walk, and discovered a small community hilltop garden with blooming wildflowers and a stone labyrinth at its center. Ha! Caught and looped in my own preaching. I took off my shoes and walked the labyrinth: inward-outward. Coyotes well deserve their reputation as sacred tricksters. If stuck in the rut of your old route with no coyote coming for you, then make one up. *Follow her.*

Wired to Wander

In 2014, the Nobel Prize in Medicine was awarded to John O'Keefe and his colleagues for their research into place and location-grid neurons residing in the hippocampus of humans and other animals (from birds to lab rats). We already knew that the hippocampus is enlarged in bird species like chickadees, who rely on the caching of food and must find their stores again in the future. O'Keefe found that location-grid neurons intertwine with neurons that involve memory, and affirm the close relationship in humans between a sense of place in space and the memory-based knowledge that runs through so many Indigenous cultures, particularly those whose lives depend on nomadic roaming.

The other day, a friend asked me how my new chinchilla was doing. A few days prior, she'd sent me a link to a supercute baby chinchilla livecam, and I didn't even wonder why—who wouldn't want to watch baby chinchillas all day? But now with her question I realized there was some confusion. "I don't have a chinchilla," I told my friend. "Yes, you do," she insisted, reminding me of various details I'd related about my alleged pet's adorableness. This was odd, but I was almost certain I was right; not remembering I had a chinchilla would be worrisome indeed. We quickly came to the captivating bottom of the muddle: she'd been walking with a different friend on the exact same path that the two of us often walk, and *that* friend had told her about *his* baby chinchilla. We stared at each other for a moment, riveted by the fascination of it. This was an actual place-based memory.

In modern urban humans, location-grid neurons are flabby and underutilized, especially with Siri displacing any wayfinding skills we might have managed to maintain. But O'Keefe's research is a reminder that knowing ourselves in relationship to place is an innate biological capacity. English writer Tristan Gooley penned a brilliant book called *The Lost Art of Reading Nature's Signs*. This art encompasses the learned craft of map-and-compass orienteering as well as the internal craft of knowing our place in relation to waterways, pathways, and the movements of stars. When we allow ourselves greater freedom in space and place than has come to be the norm, we create our own pathways of meaning and knowledge upon the land where we dwell. Wandering freely, we garner landmarks, presences, ecological awareness, a sense of kithship. Our brains and our hearts alike gather this knowledge as we become intimate with the paths that speak to us most strongly. Our footsteps in the outer world create an inner, wilder cartography that whispers, *This way, this way...*

The Wandering Mind

Mindfulness practice was popularly introduced to the West by the spare yet profound writings of beloved Vietnamese Buddhist monk Thich Nhat Hanh, who speaks to a quieting of the mind both through sitting meditation and our everyday chores, drawing all focus gently to the task at hand, whether that is engineering the design of a bridge, washing the dishes, or simply breathing in and out. In the decades since Nhat Hanh's popular book *The Miracle of Mindfulness* made its way around the world, his call to fundamental attentiveness has been studied and adopted by clinical psychology. Mindfulness-based stress reduction, or MBSR, is a practical wellness movement pioneered by American psychologist and molecular biologist Jon Kabat-Zinn, who studied with Nhat Hanh and developed meditative practices to decrease anxiety, depression, and even physical pain.

In newer studies, Kabat-Zinn and other clinical psychologists are turning the question around: what if, instead of working to focus on the present moment, it is just as mindful to follow the mind where it wants to go, to let it wander? Kabat-Zinn adopted Krishnamurti's phrase *choiceless awareness* to describe this more meandering meditation. The practitioner is encouraged to follow her distractions during meditation and so, ironically, not become distracted by them. Instead of intense focus, aimless wandering of both mind and body allow a renewed sense of calm responsiveness to our lives and world. "We inhabit an intrinsic completeness," as Kabat-Zinn puts it, "transformed throughout our nervous system and right down to our atoms and molecules." He is quick to point out that individual health is a fine benefit but not the ultimate goal. In the tangle of what Nhat Hanh calls *interbeing,* such practice is mutually enhancing to self and to earth:

> We're doing it to live life as if it really matters—not merely for the individual body but for what I call the body politic. The entire species needs to come to its senses. The well-being not only of our species but of all species really hangs in the balance. If not our calling, at least this is our opportunity. And it's to be squandered at our peril.

Bodily wandering is the same. Every spiritual tradition has embraced the significance of pilgrimage—following intuitive pathways to sacred sites, even though in ancient times such journeys were fraught with risk. As Lao-tzu taught, the journey itself carries as much grace for pilgrim-wanderers as their arrival at any particular place. The little-known fourth-century pilgrim

Egeria left a tattered diary of her wanderings to Mount Sinai, Mount Nebo, Rachel's Well, and other sites, describing pathways, orchards, mountains, gardens, and springs. Divinity scholar Robert Ellsberg writes that journeys such as Egeria's are acts of devotion that reach far beyond the personal, being "a service to those who, through her witness, participated imaginatively in her pilgrimage." Through the lens of ecological interconnection, the unfurling intelligence of our own everyday wanderings benefits all beings.

Songlines / Selflines

Wandering may be directionless but it need not be dopey. There is no reason to move about like meditative clouds. We can leap, dance, run like a coyote, wave our arms like an unruly child, like an octopus, like a tree. We can hum, sing, scream, or dwell in shadowy silence. We can pad forward, double back. We can be home for dinner. We can forget we have a home. Our paths go their own way—a new way, a strong way.

Wandering can be a disquieting activity, given the cultural expectation to at least *appear* productive. But in her 1938 classic, *If You Want to Write*, Brenda Ueland declares aimless attention essential to good work and deep thought:

So you see the imagination needs moodling—long, inefficient, happy idling, dawdling, and puttering. These people who are always briskly doing something and as busy as waltzing mice, they have little, sharp, staccato ideas...but they have no big ideas.

Directionlessness leads to clear vision and creative flourishing. Decoupled from overt value in the usual measures, wandering is an unorthodox act, removing us from the anthropic realm of striving, judgment, and economic utility. It allows us to uncover our authentic direction—a squiggling, awkward thing at first, a back and forth curving that begins tentatively and grows more assured, more knowing, more graced. Emily Dickinson was a spiritual seeker and a religious rebel (a "pagan" and an "infidel," self-proclaimed); of her many names for God, my favorite is Vagabond.

Wandering brings us to the singular songlines, the ley lines, within the soil where we live. There is no way we can go on a simple ramble and pretend to forge anything like the depth of Indigenous Australian people's true songlines, which follow the journeys of ancestral spirits, bearing profound insight into the land, plants, animals, and lore that is thousands of years old. But we can glean a sense of what a songline might mean for us in our day—a response to the sacred paths that we alone are summoned to trace. The voice that calls or hums or rustles the light hair on our arms. *This way, this way...*

Wandering for Beginners

Wandering needs no label—no benediction from economic or health or traditional religious systems (though I definitely think we should keep moodling). It is radically and essentially countercultural. It belongs to us, our mind-set, our way in the world, our freedom, our feet. Our pathway taken or not, our minds open or agitated. Our ranging limbs that take us everywhere or nowhere.

Take ten minutes, half an hour, two hours, a day. Smile lightly, like the Mona Lisa. Close your eyes, place your hand over your heart for a moment, and breathe. Let go of schedule-driven entanglements; they will be there for you when you return from your wander. This is a bit like the practice of ending every yoga practice with *savasana,* or corpse pose—a reminder that the world can do very well without us for a short time, and will eventually do very well without us altogether.

If you really do have only ten minutes, perhaps all you can do is wander around your house—visit the shining dust, the spiders who have set up shop. Follow your senses, your wonder, your instinct, your joy, the "ear of your heart," as Saint Scholastica of Nursia put it twelve hundred years ago in a critique of her famous twin brother Benedict's limiting rationality. Let this heart-ear draw you, untethered, from place to place. If you have more time, go outside. If there is no park, no grass, no woodland, then wander onto (and off of) your urban sidewalk, ending up—where? A book-lined library, or an odd little coffee shop, or a hidden garden, or nowhere at all.

We gather ourselves for the time when we must go forward—a time, indeed, that leavens our wandering time. But our pilgrim wanderings feed, rather than distract from, the sources of our resilience and resistance. We hear the call of the whales, of the oaks, of the salmon, of the cedars, of the wolves. *Walk! Sit! Run! Be still! Scream! Run! Run! Run!* There are times when we need to do all of these things. Yet there are times when wandering is right in itself, goalless, even as it tills us for a broader earthen calling.

WALK A NEW WAY

Immerse

Staying Dirty During a Forest Bath

When I began composing this book, I took several weeks away from my home ecosystem and decamped to a small cabin on UC Santa Barbara's Sedgwick Reserve, in the remote Santa Ynez Valley of central California. I believed that immersing myself in a landscape so different from my own would influence my thinking and writing, and it did— in entirely unforeseen ways.

One day I hit my usual post-lunch writing lull and decided to do what I usually do at such moments: go for a long walk. Sedgwick is six thousand acres of rare oak savannah, peopled by the twisted shapes of the live oak, blue oak, valley oak, and, while I was there, a superbloom of wildflowers during a spring punctuated by lavish rainfall, bringing a welcome reprieve from nearly a decade of drought. The air was warm, the light intense, the ground covered with grasses and blossoms that would make Red's Wolf proud.

Sedgwick is also populated by more rattlesnakes per acre than anywhere in the entire Santa Barbara region. In early spring, the snakes were waking up from their winter torpor and on the move. I was told that when UCSB herpetologists came to do a survey the previous year, they'd quickly found eighteen snakes right in the grasses and woodpiles outside my cabin door. That seemed like rather a bit many. While most people who live in rattlesnake territory rarely see snakes ("I've seen only one in my life," a friend born and raised in rattler country told me), I saw five in my short time here (not including the three dead ones). I almost stepped on two.

Early spring in this landscape also serves up an abundance of ticks in their tiny nymph phase—the stage most likely to infect a host with Lyme disease. I learned that "questing" is the verb for their act of waiting at the tips of grasses and chaparral for a host such as myself to leap upon—a lyrical word for such a worrisome activity.

During my two-hour walk, I was vigilant about staying out of the snaky tall grass and watched where I trod for sunning snake-shapes. Upon return I checked every inch of my body for the nearly invisible tick nymphs. Even so, I felt renewed, relaxed, alert, and filled with a reverence for this place that deepened each day I was there. The brisk attentiveness required to take a long walk at Sedgwick left me both calm and exhilarated.

Forest Prescribed

Wandering the Sedgwick landscape, I found myself pondering the science-based forest bathing movement. It is a poetic pairing of words—forest bath, a literal translation of the Japanese *shinrin-yoku*. The origins of modern research on forest bathing lie in the medical establishment, and many of us have been delighted to find the sensual awakening, wonder, and peace we find in the natural world affirmed by good science. Yet there is something about the forest bathing trend that leaves me vaguely disquieted, and it took me some time to understand the source of my uneasiness.

In the early 1980s, health practitioners in Japan began intuitively prescribing meditative walks in nature to patients with anxiety or stress-related physical symptoms. They recognized the benefits of "taking in the forest atmosphere through all of our senses." When the broader medical community wanted a scientific basis for the prescriptions, a host of researchers supporting evidence-based medicine took up the cause. Professor Yoshifumi Miyazaki of Chiba University in Japan is among the first and best known.

When he was nine years old, Miyazaki moved from an apartment to a house. There he experienced real soil for the first time, and spent all of his spare hours in the garden, working alongside his father, who delighted in sharing a love of botany with his son. Even as a child, Miyazaki recognized the solace that time in nature brought him. A scientist at heart, he channeled his curiosity about the connection between a calm spirit and time outdoors into his adult research.

Miyazaki notes that "humans have evolved into what they are today after the passage of 6 to 7 million years... less than 0.01% of our species' history has been spent in modern surroundings. Humans have spent over 99.99% of their time living in the natural environment." All of our psychophysiological functions are adapted to natural outdoor settings—ones that are not free of stress or danger, but with which our bodies are entangled in a symbiotic history; modern humans are responding to the evolutionarily sudden immersion in artificial and urbanized environments by operating in a constant "stress state," says Miyazaki. The biological functioning of our bodies and minds has not caught up with a culture that keeps the earthen vessels of our human forms away from the clay in which we are rooted. This makes so much sense. It is no wonder we are such a mess.

In 2016, Miyazaki and his colleagues compiled the most recent peer-reviewed scientific literature on the subject into one long paper proving that even short periods of viewing or walking in natural places results in a swarm of health benefits, including: lower blood pressure, decreased heart rate, reduced cortisol production, balance of activity and hemoglobin in the prefrontal cortex (having to do with emotional well-being and control of emotion-based impulses), improved blood glucose levels, higher immune function, and overall physiological relaxation. Though there is no evidence that links forest bathing to prevention or improvement of specific diseases, we know that relaxation and immune-function recovery have what Miyazaki calls an overall "preventive medical effect."

In a typical study, nearly three hundred participants were divided into small groups who visited both forest and urban environments. They viewed, and then walked around, each area for about fifteen minutes, and afterward their blood pressure, parasympathetic nervous activity, and cortisol levels were measured. All of these metrics showed significant improvement after even this very limited forest exposure, with more dramatic results after walking rather than simply viewing. Forest walkers had a 102 percent increase in calming parasympathetic nervous activity—*after just fifteen minutes.*

Forest Inspirited / Forest Commodified

Now that the research has been widely translated and has hit the US, we are busy doing what we do best—turning the forest bath into an industry. Organizations now exist where one can, for a fee, take classes in order to become an official forest-bathing guide or attain forest therapist certification. One popular program advertises: "The forest is the therapist. The guide opens the doors." In a typical guided forest bath, participants are led for a fee on meditative walks that last an hour, or two, or three upon gentle paths in forests, woodlands, or even urban parks, with suggestions along the way for remaining mindfully attuned to the varieties of sensory input—fragrance, soundscape, foot upon soil, wind upon skin.

Sensory embodiment is a rich and intrinsic form of wild apprehension and intelligence, a blessing upon our minds, bodies, spirits. Guided walks such as these offer a toe-dipping for people who might be hesitant about entering the woods on their own or are wary of nature altogether. Many people don't even know where to start in accessing the beyond-pavement world, and in this a good forest therapist might be invaluable. Yet it is essential that we remain mindful of how the science of forest bathing finds its way into our lives and our attitudes about the natural world.

The Japanese practitioners who originally suggested *shinrin-yoku* to their patients in a medical context were drawing on intrinsic cultural knowledge rooted in Shintoism. Though often called the Indigenous religion of Japan, Shinto predates organized religious sensibility and is better described as an ancient Japanese worldview—a way of being upon earth. At Shinto's center lies a radical animism, a knowledge that the things of the world surrounding us possess *kami*—their own form of aliveness, of consciousness, of inspiritedness. All things are aware, expressive, engaged with the world, each in its unique way.

Such animistic sensibility is not unique to Japan but has been expressed in myriad ways through varied cultures and traditions, both ancient and modern. David Abram writes in his essay "Magic and the Machine," which appeared in *Emergence Magazine*:

> For most traditionally oral, indigenous cultures that we know of, any and *every* phenomenon is potentially animate; everything moves. All things are felt to have their own pulse, their own inner spontaneity or dynamism. All things have agency, the capacity to act—although some things, like trees, rocks, or mountains, clearly move much slower than other things, like bears or dragonflies.

One summer day I was straggling with several other white Seattleites at the edge of the Indigenous tent at our city's annual Folklife Festival, and the Haida teacher-storyteller who had the mic waved us in. "We are all one canoe. Come in, come in." I was raised in a culture in which allusions to anthropomorphism and animal consciousness have been thoroughly expelled from scientific discussion, and I felt moved by the forthrightness and absolute lack of any need for justification with which this teacher said, "Spirit. All things have a spirit. We know it and it is true. We share this with you because we love the children. Teach this to your children." *We know it and it is true.* To live rooted is to stand in the courage of such conviction.

One of the things that makes Japanese animism unique is its collision with Buddhism, which originated in India and made its way via Korea to Japan in about 550 CE.* When Buddhism's contemplative stance came under the influence of Shinto's deep animism (and vice versa), an authentic *shinrin-yoku* arose: quiet walking, with bare feet or handmade *waraji* sandals, under the influence of a vividly inspirited world that is likewise conscious of human presence. This is a transpersonal alertness, what Celtic philosopher John O'Donohue calls "interflow"—a reciprocity of life, of pulse, of natural wisdom, an offering and receiving in equal measure, and without question.

The foundation of modern forest bathing is different. A journalist writing in the *Atlantic* states the goal of forest bathing clearly as a "means to an end, and that end is better health." *Human* health, that is. The health of the land, trees, creatures, and ecosystems is rarely mentioned in the nature-healing literature. Though it may inspire an interest in nature study,

---

* Since a codification of Shinto beliefs was undertaken beginning in the 800s CE, it sometimes seems on paper that it grew alongside Buddhism in Japan. But Shinto existed as a worldview and a way of being long before the arrival of Buddhism.

knowledge of the species and ecological relationships of a particular place is not part of the forest bath paradigm. Stripped of the pretty pictures and poetic language, the *business* of forest bathing frames nature as a commodity for human use—a pharmaceutical-like prescription, a day spa, a therapist's chair. Studies pinpoint the least amount of time in nature that is "enough" to allow people to reap physical and mental health benefits—it turns out that the "minimum dose" is twenty-five minutes outdoors each day. As a health resource, forests themselves can be made to fit into a modern, human-centric reckoning in which wild creatures, places, and even cultures must prove social and economic value to gain attention and support. It is more than mere semantics to recognize that while time in the forest is therapeutic, the forest is not a therapist. Not a pill, not a spa. Not a cog in the capitalist wellness industry.

The science behind forest bathing is beautiful, *completely beautiful*, affirming an elemental connection between our bodies and the body of the planet. Guided forest baths can inspire wonder and love. Yet we must remain alert to the ways we engage this science. The line between commodity and reciprocity may seem thin, but it is profoundly meaningful.

The Replaceable Wild

Much of the research on forest bathing and nature healing is leading to explorations of how the benefits of the natural world for humans can be harnessed without any actual contact with a real forest, a single tree, or even outdoor air. Researchers extrap-

olating from nature healing have learned that prisoners in solitary confinement exhibit less stress when they work out in a
room showing videos of natural scenes. The simple fragrance of
certain trees invokes calm, whether the trees are there or not, as
do recorded nature sounds, and even gazing at a houseplant.

These are meaningful positives. Nature-based medicinal practices are a basic way of relating to the earth—ancestral ways
made new—and may replace more invasive pharmaceutical treat-
ments. When I am not writing outdoors, I sit at a desk piled with
stones, shells, plants, fir cones, and the fragrance of cedar essential oil wafting in the air. Yet there is a slippery slope where easily
accessed "nature-derived stimuli," can come to replace our necessary attentiveness to living nature. This imitative wild is harmless
enough if we recognize a poor substitute when we see one, but so
often we don't, especially when—as in the case of wild nature—
so many of us modern humans know very little of it to start with.

In an ironic (and disturbing) turnabout, research on the
health benefits of easily accessible nature may put remote
wildlands—those beyond roads and trails and human utility
that must be strictly protected in a time of mass extinctions—in
danger of seeming superfluous. As I write, *Walden: A Game* is
getting rave reviews. In the game's dazzling digitized scenery we
can listen to birds, avoid poison oak, and ponder the meaning of
life inside Thoreau's virtual cabin, all without leaving our screen
or experiencing any visceral-olfactory input at all. The game is
being heralded as "world-changing." We don't even have to wonder what the man himself would have thought. As Jack Turner
writes in his passionate, philosophical book *The Abstract Wild,*
"intimacy with the fake will not save the real."

One researcher, engaged with the science of forest bathing, proclaims, "Without it, we'd be back at Walden Pond." He refers to the actual Walden Pond—not the virtual one—in his thinking a step backward from our modern scientific understanding of the human experience in nature. Yet to my mind this poetic research does indeed, and thankfully, send us back to Walden Pond or any immersion in the earthen wild, where Thoreau went to ask not "What can nature do for me?" but *How should I live?*

The Calm Anxiety of Creation

If you picture me serenely writing this book with a pencil in hand and a rustic notebook in my lap while basking in dappled sun beneath a leafy old moss-and-fern-covered maple, you would be, much of the time, mostly correct. I always prefer to write outdoors, in the path of my subject matter, and I contrive all kinds of ways to make this happen, even when the weather is not summery. I wear fingerless gloves in the autumn, carry a thermos of spiced tea in search of trailside log-desks, light a fire in the backyard fire ring and pull a chair up close. When it really is too cold to work outdoors, I still take long walks, and try to carry the rooted sensibility I gather back to my studio, the window open just a bit so the outside can find a way in.

But the "serene" part? Anyone who knows me would laugh out loud. Sometimes, sure, I sit there calm and sweet among the trees in classic nature-writer pose, face upturned to the voices of birds like Keats under his plum tree, twirling my whittle-pointed pencil. In truth, though, I am overall a fretful writer, plagued by anxiety and insomnia and self-doubt and deadline dread and procrastination and self-recrimination for my procrastination. Many days I would trade the blank page for a barefoot walk in a field of riled rattlesnakes—and gladly.

When I get super stressed, a walk in the woods is the best curative. Poets and scientists (and poet-scientists) often suggest that the resulting calm comes in large part from the *tranquility* of the natural landscape—the soft focus we receive in the presence of lapping waves, whispering leaves, wandering clouds. But, while I'm not finding any science to support it, I have come to realize that it is not only the peace of the forest green that settles my overwrought nervous system. Quite the opposite: it is the simultaneous recognition that all organisms in the natural world are just as anxious and vulnerable as I am.

The lives of most creatures, except the highest-level carnivores, are spent in a state of constant vigilance to avoid being eaten. Plants, too, share this evolutionary compulsion, spending bioenergy in the production of thorns, toxins, warning color, and chemical communication with their botanical kindred to avoid the nibbling of insects and deer. Wolves, cougars, and grizzlies, meanwhile, face the cruel threat of not being able to provide enough meat to sustain themselves and their young. The natural world is a severe, uneasy place; the depressive anxiety that visits me sometimes finds purchase there, and consolation—not in calm beauty, but in continuity, in mutual empathy. We are all in this together. Alert after a walk in my Northwest woods, I look up to see the talons of the Cooper's hawk hot with the blood of the Swainson's thrush, there at the forest edge; I am enfolded, seen, wildly serene. I steady my shaky hand with a deep breath of tree-air and lift my pencil again.

Staying Dirty During a Forest Bath

Even when forest bathing involves an actual wooded path, the experience is often ironically dewilded, and the challenging truth of the forest minimized. The experience reported in the *Atlantic* is illustrative: "Our guide had already tackled the hard part of finding a trail with minimal elevation gain and limited poison oak along its flanks." True, the goal of the forest bath is inner peace, holistic health, and mindfulness through time in nature, not an actual wilderness experience. But I wonder if one can have the former, truly, in all its depth and breadth, without the latter. The fruitful empathy of the fretful, complex wild must be allowed to touch us. I think of the stinging of nettles that so often brush my hands as a reminder of the woodland's beyond-human aliveness and my own need for attentiveness—as effective as a meditation bell. I think of the words from *Benedicto*, by Edward Abbey, which my sister read at my wedding and I copied into my daughter's graduation card: *May all your trails be crooked...*

And I think again of the rattlesnake I almost stepped on—my pumping blood, my fear and adrenaline spike, my giddy laughter when I realized I was safe, my brightened awareness of all the footsteps that followed, my visceral *intimacy* with the land and all that is on, above, and below it. Including the snake who didn't want to bite me anymore than I wanted to be bitten, both of us standing so close together in this brisk, uneasy calculus of the wild.

Walking in the woods in order to de-stress and seek peace and be in love with the earth is sacred time. This is what most of us mean when we speak colloquially of forest bathing. And certainly the human-health-based emphasis in forest bathing is harmless next to other, more impactful uses of natural places. Still. My wish for all of us is forest baths guided by our inner knowing. Where we don't have any idea what is behind each turn. Immersed, unsettled forest baths—ones where we emerge with ankles enlivened by the prick of nettles, lichens in our tresses, pebbles in our pockets, an uncertainty about whether the tendrilly growth on our arms is hair or fur. Our heart rate calm yet beautifully feral. Let us return so mingled with the stuff of the earth that the first person to stumble upon us after we are home from our wandering looks at us and says, with a mixture of admonition, admiration, worry, and an urge to suddenly run out the door themselves, "You need a bath."

DECOMMODIFY THE FOREST

Alone

The Essential Complexity of Solitude

My cute mom, Irene, called me the other day, as she does periodically, to voice her worry over my love of camping and hiking by myself; she knew I was plotting an escape in the near future. My mother loves me, but I seriously believe she would have me committed to an institution just to make sure I don't fall off a trail ledge somewhere, or get attacked by the bear of her maternal imagination. I have been going off alone with a pack on my back since high school days, and I ask her why she didn't oppose me so vocally back then. She says, "You wouldn't listen to me." It's sweet that she thinks as a grown woman with my own grown daughter I will listen any better now. Of course we must take all commonsense precautions to keep safe, but in my own experience nothing leads more swiftly or thoroughly to a deepened sense of wholeness, an apprehension of interflow with all of life, and a clarity about the work I am called to bring to this moment on earth than periods of extended solitude.

A Culture of Unsolitude

Time alone in nature is essential, but our culture is not set up for such interludes, and most modern humans are unaccustomed to and ill prepared for them. Now even the little time we do find to be alone with our thoughts—waking early or staying up late to have the household to ourselves, even riding the bus or walking in the city park—is interrupted by the constant technological intrusion of alerts from the phones we carry always. If we become uncomfortable in our moments of solitude, it is easy to find a hit of distraction on social media, that swift shot of yummy dopamine, making us all unwittingly addicted to such activity, even as we judge much of it to be insubstantial and disruptive.

In a widely covered study published in the July 2014 issue of *Science*, Timothy Wilson of the University of Virginia and his colleagues provided participants with electroshock devices that they could use on themselves entirely by choice, and were asked to be alone with their thoughts for just fifteen minutes, with no distraction other than their option to use the shocking device. The subjects all tried the device out before the fifteen minutes began so they would know the exact nature of the bargain, and though many of them said that they would actually pay money to avoid the shock, when confronted with the brief period of undistractedness, one quarter of women and two thirds of men chose to give themselves the painful shock rather than endure the simple stillness. One subject shocked himself 190 times in fifteen minutes (let's call him an outlier). When different subjects were tasked with taking between just six and fifteen minutes of undistracted time alone at home, over half admitted to cheating— turning on music, checking their phones, flipping through a book.

Safe in the Woods

My husband, Tom, has given me the great gift of respecting my need for times of prolonged solitude. As soon as our daughter, Claire, was weaned from breastfeeding, Tom gave me a present— a night away in a village town on the Salish Sea. As Claire grew, he always supported my going away for longer periods to think, write, or simply be alone, which I consider, also, a gift to our daughter—the example of mothers having a valid creative life and time for themselves.

The question people most often ask about my experiences in solitude is not "What did you see?" or "What did you learn?" but "Weren't you afraid?" And while I am more eager to answer the first two questions, the truth is that I *am* often afraid. When I sleep in the woods alone, I am almost always scared out of my wits. Years ago I did a solo alpine backpack on Mount Rainier, guardian of the Cascadia bioregion, called Tahoma or Tackoma by Coast Salish peoples. The climb was long and hot, and I was exhausted by the time I reached the small shimmering lake that was my destination. Dusk fell with grace upon my mind and body as I sipped my soup and watched the shadows overtake the water. I knew I would sleep as sound as a black bear cub. Which did not happen.

As soon as I zipped into my dark little tent, the night sounds, real and imagined, began. Shifting leaves, sighing trees. A snapping twig. My screaming heartbeats. Footsteps falling.

From what? A mouse. A bobcat? A man, oh god, it is a man. How did a man get here? Did he watch me eating soup? No, there is no man. But I heard a growl. Wait, did I? I did. Something big. A cougar. I am going to be eaten by a cougar. And he will drag me away and there will be nothing but my diary and some blood left at the campsite and my bereft family will search for my body forever, and just find . . .

I lay awake, thinking up ever more grisly details of my demise, berating myself while plotting my escape.

Why did I come here? Why did I do this? My mother was right. I am so stupid. I'll leave first thing in the morning, eat breakfast farther down the trail just to get away from this creepy place. I can start packing now, the small things in my reach. First light, I'll get out, get out. And next time I think of hiking by myself I will remember this and NOT do it. I am never, never doing this again.

I must have fallen asleep at some point. In the morning, I woke up, listening for twigs and winds and growls. There was only quiet. I unzipped the tent flap and peeked beyond it. Mist rose from the still lake. I crawled out, stretched. The alpine firs were already fragrant in the morning light. On the water a pair of hooded mergansers were gliding with five downy merganserlets. And across the tiny lake? A small herd of elk. Velvety, white-rumped, regal, soft. Their hoof-steps chimed through the mist. So near.

What was wrong with me last night? Damn, I was nutso. How did I let my imagination run away like that? Look how beautiful and safe this is. Oh, what a gift to be here alone! I am

SO glad I didn't run off in the morning dark—such peace I would be missing. That would have been so stupid!

And when night fell it started all over again.

I always thought this was my private little misery, but when I started talking to girlfriends who are solo hikers or campers, I would hear my ordeals echoed exactly. Jen Rose Smith covered this subject well for *Outside*. She would simply lie awake in her campsite from darkfall to sunrise, making herself exhausted enough to "sleep through a horror film marathon" by night three. Other experienced and intrepid women hikers Smith interviewed told her that they face "a nighttime head trip while alone in the tent," paralyzed with fright and picturing gruesome newspaper headlines. One woman said, "There's a certain point in my fear where I'm so scared that I can't talk myself out of it and there's no reasoning with my brain." Crazy night-thoughts seem to come with the territory.

But while anyone might lie awake fearing the attack of a bear, a lightning strike during a storm, or a tree felled upon their tent in the wind, it is most often women who fear other humans and men in particular. Our fears, writes Smith, "circle around people and contain an unmistakable sexual twinge." She notes that growing up as a woman involves constant warnings about going out alone, and reminders of the precautions we must take: walk in groups, don't go too far, maybe carry a weapon, certainly carry a cell phone. I remember Little Red: Stay on the path; go straight to your grandmother's house. When Smith shares her love of solo camping with others, like me she is met with a flurry of "well-intentioned advice," the most common being to bring a friend.

This is not poor counsel.* It is surely sometimes safer to hike with another person. But that doesn't mean hiking or camping alone is *un*safe—or at least not less safe than many other things we do that are socially accepted. Smith turns to the numbers: In 2014, there were eighty-three rapes on National Park Service land (that's one in 3,527,951 visitors) versus 84,041 (one in 3,794 people) in the country as a whole. Also in 2014, there were 14,249 murders in the country and only sixteen on Park Service lands. And while people we meet on the trail are likely all strangers, over 75 percent of women who are sexually assaulted know their attackers. Smith writes, "Like a person who fears airplanes and sharks over highways and heart disease, my nervous nights in the woods just don't reflect the world's real danger."

I was delighted to find this outlook echoed in an 1862 letter from Emily Dickinson's hand, when she was thirty-one. The poet is mythologized for her later reclusive nature but roamed wildly into young womanhood: "When much in the Woods as a little Girl, I was told that the Snake would bite me, that I might pick a poisonous flower, or Goblins kidnap me, but I went along and met no one but Angels, who were shyer of me, than I could be of them." This is not to say there is no risk. But the sacred rewards of wilderness solitude, for me and for many, are worth

---

* It is, of course, the advice I give my own daughter, even while wanting her to know the experience and courage that wilderness solitude brings. But this very day she came home from a solo ramble in the Cascades and said, "Now I see why you don't want me to hike alone." My mother-mind jumped immediately to: "Why? Did someone frighten you?" She said, "No! Because you knew that once I started hiking alone I would know how glorious it is and never want to do anything else ever again!"

meeting both the challenge of the night-crazies and the calculated risk of engaging any demanding activity.

The Primal Sanity of Solitude

The desire for solitude is acute—studies find that 85 percent of us are craving time alone. Yet most of these same people also fear such time. Psychologist Bill Plotkin writes in *Soulcraft: Crossing into the Mysteries of Nature and Psyche,* "Maybe we're afraid of solitude because it threatens us with boredom or an anxiety that can lead to difficult truths, unfinished emotional business, and the shadow side of human nature." I am rarely bored. But the part about "difficult truths" speaks squarely to my own experience.

When I venture into wild places for periods of solitude, it always takes me a while to sink into the presence of little but my own mind and its experience of the world, no matter how beautiful a place I find myself in. The city and the trappings of everyday life cling to my skin and my brain, and shed themselves with reluctance. But after settling in, I find myself serene for a time, a lioness in the sun, basking in the sweet, rare space, my thoughts roaming wild, my connection with the life around me bearing peace and enchantment.

Within hours or days, this sense of recollection is often displaced by a teeming, shivering anxiety. All of my schemes and plans call to me. Solitude mocks them all, and me along with it. I get edgy and fretful. My thoughts become anything but sweet as all of my insecurities, worries, and secret fears are laid bare and scrutinized: those about myself, my loved ones, and the earth. I can't make it stop. I worry about my coherence of mind: *Maybe this time it won't come back?* Sometimes my fancies seem to take form and rise as voices, as seeming physical presences.

Once a wolf prowled the edges of my camp for days. I had set up in the westernmost region that gray wolves have wandered into since the beginning of their Washington State recovery.* There were three in the area, rarely spotted. I longed to see one but knew my chances were infinitesimal—so what was this lupine visitor? I couldn't *quite* see him, yet he wouldn't leave. "Be still," my mind whispered, then screamed, "He does not exist!" Until finally I did become still, accepting near-glimpses, expecting no more. I watched the stars that night, and slept, and dreamed. The next morning I woke to tracks around the fire ring, and an oval depression where a creature had curled at the bottom of my bed, ashimmer with gold-gray-brown-black fur. Was this a dream? No. Would anyone else be able to see the fur? I don't know.

I remember Walt Whitman's invocation in *Leaves of Grass:* "Give me solitude, give me Nature. Give me again O Nature your primal sanities!" I remember Emily Dickinson's "Divine Insanity," also founded in observation of nature. Societally condoned sanity is not the only—nor the highest—sort.†

---

* Gray wolves were extirpated from Washington State in the early 1900s, but in 2008 began to cross the border of British Columbia back to the northeastern corner of the state; thus we refer to "wolf recovery" here, rather than "wolf reintroduction," as is common elsewhere. The gray wolf population continues to slowly increase and spread in spite of the Fish and Wildlife Department's aggressively killing the endangered wolves to appease farmers who graze cattle and sheep on public land.

† In various drafts of this chapter I deleted then replaced then deleted then replaced this little note about voices and presences over and over, fearing I would seem even more mentally precarious than I actually am. But I left it in

I've been through the cycle enough times to know what happens after the period of tears, the downward spiral, and the temptation to flee. After that there is nothing. A quiet mind. Darkness that reveals starlight, a blanket of strange comfort. Birdsong that is bright and vivid and speaks my truest name. Every cell of my body sensing and unbound by the flimsy husk of skin. Psychic layers shed and wrung and replaced by a deepened sense of my inner being, of my place on earth, of interflow, of my simultaneous aloneness and enfoldment in the natural world. A crazy hermit's laughter. Good crazy.

The Solitary Brain

Mystics and poets regularly laud the rigorous self-examination that solitude requires. But I was surprised to discover my experience mirrored so precisely in scientific research regarding the effects of solitude on the brain. Dr. Marcus Raichle, a professor of radiology and neurology at Washington University, found that our brains are at most times incredibly full, active, and "on." We've heard over and over that we are using only a small

---

the end because it is my true experience, and also because these very kinds of visitations of voice and vision are one of the reasons that humans for centuries have sought wild solitude. Yet there is a fine line here that I am not in the least qualified to walk on others' behalf: hearing voices can be a sign of schizophrenia and other serious mental health conditions that require professional counsel and care. Anyone in a period of major depression, anxiety, regular panic attacks, or mania, or who is struggling with addiction, should seek sturdy personal footing before venturing into extended solitude.

portion of our brain at any given moment, but according to Raichle, this is simply not true. Rather, our brains are running at about 95 percent of their potential most of the time. There is a shocking amount of background activity—more than ever in this era of constant technological connection.

In the 1990s, Raichle and his colleagues were studying the parts of the brain activated by certain tasks. They unexpectedly discovered that there are a variety of areas that light up only when there are no external distractions. These include highly potent "self-referential" processes, including the recall of personal memories, emotional states and feelings, and the evaluation of sensory input. When external distractions are removed, these parts of the brain are allowed to work at full capacity.

No wonder solitude is so unnerving, powerful, and essential. Solitude lets the enormity of our brainscape roam free. Solitude allows our most sensitive feelings and memories and self-judgments to surface. Solitude frees us from what psychiatric science calls the spotlight effect: the tendency that we have in public to overestimate the attention others pay to our accomplishments, our errors, our appearance, and the words that come out of our mouths. Solitude unshackles us from the compulsion (for some, an addiction) to curate and display our lives on social media, thus allowing our interactions with ourselves, others, and the natural world to be entirely what they are in themselves, not superimposed upon an artificial narrative for which we seek validation and approval. Solitude allows our brains to form interconnected neural root strands beyond those we typically utilize.

These are among the reasons that thousands of saints, seekers, artists, philosophers, and students of nature across times and cultures have conspired to create periods of aloneness for themselves. Here there is a shedding and an emerging simplicity (time in nature shows us how little we truly need the material things with which we over-outfit our modern human lives). Here, the layers of the psyche peel away, one at a time, until we get to a difficult, shimmering center. Here, we comprehend "our unique way of belonging to this world," as Plotkin puts it, outside of the expectations of society, family, and—especially— ourselves.

In the poetic rightness of etymology, the word *alone* comes from the Middle English—not separateness, but *all + one*. Both at the same time. While extended solitude does lead to a sense of inspired ease with our individual mind, it also leads us paradoxically into a deeper comprehension of our essential interconnection with other humans and the beyond-human world. We look around us at the trees, the birds, the flowers, the foxes and find they are not "other" at all.

Yet not all aloneness is equal. It is important to distinguish between positive solitude and unwanted isolation. To be beneficial, solitude must be willingly chosen, and there has to be an out—the option of returning to societal life if needed or wanted. The effects of chronic social isolation are grim, including a 26 percent higher mortality rate for people isolated from human company due to life circumstances than those who live with a modicum of human company.

The repercussions of isolation for prisoners is horrific and sometimes irreparable. Compared to the general prison population, isolated prisoners develop psychiatric disorders, physical health decline, diseases without overt cause, and death at a radically higher rate. Intentional, positive solitude has none of these effects. Poet Jane Hirshfield writes in "Vinegar and Oil," from her exquisite collection *Come, Thief:*

> *Wrong solitude vinegars the soul,*
> *right solitude oils it.*

Flourishing

Solitude is at times an extreme challenge—that is one of the reasons we seek it, part of its wild grace. It can also be an economic privilege, difficult to find as modern workers, caregivers, students, and parents. We steal moments as we can, and seek longer spans as we are able. A solitary retreat may be, at first, only an hour here and there, in the backyard, or beneath a park tree, or in the middle of the night.

Leaving behind technology during intentional time alone is essential for the cognitive benefits, neurological repair, and spiritual clarity that are the gifts of solitude. Multiple studies show that anxiety is markedly reduced, and we gain benefits similar to solitude, not by simply turning our phones off but by having them *not physically with us.* If a phone is essential for safety during time alone, then turn off alerts, cover the screen—just tape paper right over it—and keep it somewhere that is terribly inconvenient to access. I am always surprised by how long it takes me to give up the impulse to reach for my phone, often for no reason at all, other than to "just check." Check what? Always it is something that can do without me for the moment. It is imperative that we allow ourselves time to free our minds from even the possibility of constant connectivity, to "normalize deactivation," as herbalist Sophia Rose puts it, allowing our enchanted neuronal connections to rest and reassemble.

Consider letting go of time altogether — give up reference to linear, numbered hours in order to apprehend the rhythm of the natural world and the needs of your own body. Sleep when tired, eat when hungry, and be awake whenever you want — the middle of the night is not just for nightmares and insomnia, but for night sounds and moonlight and a mind undistracted by the sunlit shapes of day. Bring little in the way of distraction, perhaps a notebook for exploring thoughts, but maybe not a book. It is our own minds and visions we seek, not those of others. When anxiety and doubts feel overwhelming, try letting them wash over — there is a bottom, and clarity on the other side. Psychic shedding and the resulting mycelial tendrils of connection that form within our beautiful brains are among solitude's dearest benefits.

Freaking out? Talk yourself down. Remember that you came here on purpose, even knowing it would be difficult. Remember the vision that you had for yourself in solitude — calm, alert, open. Courageous. Remember that human life, any life, is not always blanketed by ease and comfort, that our minds are conditioned to be anxious in new situations, and here you are, a little uncomfortable but entirely safe. Focus on gratitude for this time and remind yourself again: *You came here on purpose.* Anything and everything you are going through has been shared by thousands of other solitude seekers throughout time.

Return

The need for return is almost universal. Some few solitaries do find their peace and vocation in the hermit's life of contemplation, their stillness reverberating upon a rapidified earth. For the rest of us, there is a different calling. Past the fitful impulse to run from solitude and the peace that follows comes another sensibility: it is time to gather the lessons bestowed and return anew to daily life and company.

Thomas Merton discovered that the life of pure solitude he longed for ran against his human nature. He worked for years to convince his superiors at the Abbey of Gethsemani in Kentucky to free him from his regular monastic duties and communal prayers and allow him a hermitage in the woods. When his wish was finally granted, it was barely a week before he reported grumpily to his journal that none of his brothers had come to check in on him (and he never found the heart to live as a true solitary).

Some solitaries feel summoned to return. Catherine of Siena was a precocious and bookish girl. Against all convention in fourteenth-century Italy, and inciting the fury of her father, who wished her well married, Catherine locked herself into her room at age sixteen, creating her own little monastic cell. She planned to live her entire life there in asceticism, contemplation, simplicity, and sometimes ecstasy. After three years of this existence (her father finally resigning himself to it), Catherine saw a substantial vision of her divine Beloved standing just outside her doorframe and calling for her to emerge. "The service you cannot do me you must render your neighbors." The Beloved, she perceived suddenly in a vision shared by mystics across cultures, was in everyone, everything, everywhere. Catherine started with the basics, going down to family dinner, where she was greeted by her stupefied parents as well as sisters-in-law and children who had joined the clan during her solitude and whom she had never seen. She spent the rest of her life working for social justice.

For most of us, the call for home is a sweet one, and no less wild than the longing for solitude. Time alone enkindles life in the earth community.

SOMETIMES, GO ALONE

Unsee

Return to a Fruitful Darkness

I've never had good night vision. Even as a child, I noticed that I couldn't see in the dark as well as my friends could. So when I graduated from college and moved to Minnesota for a spell to work as a naturalist at an environmental learning center, I decided to find out what I could do about it. I didn't think there was a way to actually improve my night vision; I just wanted to be more comfortable with what I had—to be at home in the darkness.

I lived in a run-down but cozy trailer home with the other naturalists, and: a ferret who liked to sleep in our beds and nibble our toes; a lake virtually outside the door; loons upon this lake who would sing us awake and asleep and awake again; and the usual wildfire romances that flare and die when you throw twentysomethings together for strings of hot summer days and schedule-less nights. The ELC was at the end of a dirt road at the end of all northern Minnesota's dirt roads. Only a few miles of lake-infused earth separated us from the official Boundary Waters Wilderness, and then Quetico in Canada. Our Flathead

Lake had a three-mile trail around it—perfect, I decided, for my night-vision project.

Footsteps in the Dark

The experiment went like this: I would walk around the lake every night, alone, wearing moccasins with thin leather soles so my feet would be somewhat protected from sharp stones and errant porcupine quills but could still feel the ground. I would pick my way lightly over the soil, learning to walk at least somewhat sure-footedly without the safety net of daylight. To keep myself from cheating, I decided not to carry a flashlight, not even one with a red filter that protects night vision. (I didn't say it was my safest experiment ever; certainly I didn't tell my mother.) But unlike my solo camping trips, I was not sleeping alone out there, and after a few tentative nights I rarely felt scared. After all, I knew the path well, I was close to my funky little home, the trail was flat with nothing to fall off of, and my friends were aware of what I was up to and would seek me out if I didn't return in a reasonable period of time.

But nothing ever did happen. Oh, I tripped and fell flat on my face a few times. And my butt. I got bloody knees. Otherwise it was an innocuous inquiry, and a beautiful one. As predicted, I did not get any better at seeing in the dark, but I did get more comfortable not-seeing. I wandered with confidence and loved my walks more each night. As dusk fell, I would recall with delight the Tohono O'odham chant shared by American Buddhist roshi Joan Halifax: *Toward me the darkness comes rattling. In the great night, my heart will go out.* And though I said that overall nothing ever happened, actually one night something sort of did.

Halfway around the lake, beneath a shining quarter moon, I heard a great *kersplosh* just beyond the trail. A *very* great *kersplosh*. Even allowing for an overactive nighttime imagination, the sound could only have been made by an enormous creature. We were in wolf country, but none had been seen at the ELC for a long time. Cougars were exceedingly rare here. There was the resident beaver, friendly and blind in one eye; he liked to swim alongside our canoes but was surely snug in his lodge at this hour. We had tons of white-tailed deer, which is what I supposed this was. And we had moose.

As a Pacific Northwest girl, I was familiar with elk, but this stint in Minnesota was my first time ever encountering moose, and I instantly loved them. Huge and gangly and graceful in a strange moose-way. I was enchanted by their oddly shaped antlers and the featheriness of their eyelashes, luminous in the headlights—because that is where we almost always saw moose—standing in the middle of the road when we were on our way home from having a beer or two in Isabella, the nearest town (population at the time: 83). Standing and not moving. Taking up the whole damn road. We'd wait, we'd honk, we'd yell. But a moose doesn't move unless she wants to.

Most of the other naturalists I worked with were from the region and knew more about moose than I did, but their reports were conflicting and dubious. Some told me moose are dangerous. Some told me they are benign unless provoked or frightened. Everyone agreed that one should allow them a safe distance, just to be sure.

After the *kersplosh* there was silence, so once my nerves calmed I walked a few more steps and heard the sound again— heavy footsteps now at the shoreline, and then a silhouette to go with them—a cyclopean moose. Antlerless in the summer, this was likely a female, come to nibble the wild rice that flourished at lake's edge. And now here I was, literally five yards from a full-grown moose. "Run," my brain said. "Stay," my stupidity said—"you have never been this close to a moose, how exciting!" *Run! Stay! Run! Stay!*

Meanwhile, my new moose acquaintance was gauging the situation for herself. She stared at me for minutes while I stood there stone-quiet, my naturalist's curiosity/stupidity winning out over common sense. I waited to see what would happen next, took a few steps back to give the moose more space, and except for my pounding heart and the blood I swear I could hear rushing through my veins, I was as still as I have ever been. The moose continued out of the water, and seemed to be lurching in my direction. *Run!* started to sound more sensible.*

But then the loveliest thing happened. The moose took her too-long legs and bent them, lowering herself slowly and carefully to the ground, front first, back end laboriously following. It took forever, and was in no way a traditionally graceful effort. Eventually she managed to get her legs folded neatly beneath her chest, like an elaborate cat. She looked at me, then looked away, toward the lake, thinking her night-moose thoughts.

---

* Disclaimer: Leaving *would* have been more sensible. Never approach a moose (or elk or really any other wild creature at all). If you find yourself unwittingly in a moose's close proximity, as I did, then calmly depart.

Well, OK—if this moose was comfortable enough to sit among the trees and grasses with such a singular interloper, then so was I. I took a couple more backward steps to the other side of the trail, then sat down too.

The moose continued to contemplate the thick darkness, and I could just make out the liquid of her eyes shining. She closed her moth-lashed lids. I sat with her until I was shaking with cold, then stood up slowly. My sleepy moose, if she registered my movement, did not show any sign of it. In such unlight I could both see and not-see the moose. I had to soften my eyes, let go of my preconceived notions of sight. "Seeing in the dark" would never be a proper goal, maybe. Here was a new kind of clarity. True night vision, empty of cultural conceptions, is essentially an *unseeing*.

It seemed rude somehow, a failure of wild etiquette, to cross in front of the moose and continue on the trail. So I turned around and retraced my footsteps back home in the darkness, full of wonder.

When Darkness Became Evil

Somehow, in our language and in our psyches, we have come to equate good with light and evil with darkness. The symbolism runs deep. We see it in our poetry, our religion, our songs, and our cultural mythology. "Those in the darkness have seen a great light," the Christian hymn proclaims. In churches everywhere, we "wait in hope for the light"; in hard times we remember that light will return. People who create positive societal change are "light-bearers." In *Star Wars,* the Jedi see "darkness" in people, or that a person who has gone to the side of destruction has no "light." The entire enemy side is in fact called the dark side. For Tolkien, darkness was deployed as weaponry by, yes, the Dark Lords of Middle-earth. The "IT" in Madeleine L'Engle's *A Wrinkle in Time* is manifested in spreading darkness. Susan Cooper's young heroes fight, always, against the evil Dark. We live, we know from political and environmental pundits, in a "dark time." I could go on for pages.

Ursula Le Guin belied the penchant in literary fantasy to paint the evil as dark in her groundbreaking Earthsea books, where even the title speaks to the falseness of such duality (it is little wonder that she later composed one of the most beautiful English versions of Lao-tzu's *Tao Te Ching*, where all false dichotomies unravel). The naive young wizard Ged, in a display of ego while arguing with a rival, calls forth his own shadow and looses it in the world. He spends years (and several books) fleeing it, fearing it, trying to prevent it from destroying all that is good and fruitful in the world. But in the end it is his facing it (with the help of dragons, of course)—integrating the complexity of his shadow, his disassociated dark self—that leads to his own salvation and that of all Earthsea, a return to the balance sung in the world's Creation of Ea hymn:

> *Only in silence the word,*
> *Only in dark the light,*
> *Only in dying life*
> *Bright the hawk's flight*
> *On an empty sky.*

The Sheltering Darkness

I'd never realized how ingrained this symbolism was in my own psyche until I resolved to compose this entire book without using it. I've caught myself unwittingly plopping it in a line over and over again, then crossing it out.

It is a powerful metaphor. All of us have felt the serenity of sunlight upon our face at just the right moment, bringing warmth, healing, comfort. All of us have felt the relief of sunrise after a long night of restlessness or insomnia. All of us have found mercy in a new day, offering much-needed renewal. And yet all of these things can be embraced without banishing darkness as antithetical to the good in our modern cultural mythology.

The trees and all plants are our great mentors in this, growing, always, in two directions at once. Hildegard of Bingen grounded both her ecospirituality and her herbal medicine in this knowing. Her central concept of *viriditas*, or divine "greenness," sanctifies more than the literally green mosses and leaves; over and over Hildegard calls upon the interdependence of trees, seeds, spirit, and soil as sacred cocreators. She used various trees in her practical cures, but they were also stand-ins for the mythic tree of life in her writings and visions. In one of Hildegard's greatest hymns, she sings to the trees on her lush Rhineland. "O happy roots, mediatrix, and branches..." The "happy" roots, ever underland, grow in the particular favor of the divine (and perhaps of Hildegard herself—there is speculation that some of her visions may have been enhanced by the hallucinogenic mandrake root, which she prescribed as a cure for depression). Green branching is only and always mirrored in rooted darkness—the place we, too, come from and return to. To be whole on earth is to embrace this exuberant darkness, the landscape we visit with every turn of day into night.

Darkness possesses its own essential grace. It is darkness that bears liminal imaginings more difficult to access in the scattered daylight. Darkness brings the restorative sleep and dreaming our bodies and psyches require. Darkness takes the harried busyness of the day and transforms it to stillness, to quiet. Darkness brings us starlight. Darkness erases our view of the horizon, forcing our reliance upon a spacious inner vision that daylight cannot provide. Darkness offers a complex refuge for all beings and all aspects of being. As John O'Donohue puts it, "Trees, mountains, fields, and faces are released from the prison of shape and the burden of exposure. Each thing creeps back into its own nature within the shelter of dark."

Darkness twins with light as the primordial ground of existence. So many myths (Gilgamesh, Orpheus, Persephone, Aeneas, Dante) deploy the Greek journey of *katabasis*—descent into a dark underworld that will edify a heroic seeker's eventual life after they return to the aboveground world. Yet for the majority of organisms on earth there is no "return" to light; fullness of being is found in darkness itself. Ninety percent of life unfolds in complete and eternal darkness—beneath the soil that cradles every footstep, and beneath the sea, where whole worlds exist absent the penetration of any light at all. Without absolute darkness, the seed will not germinate, the decomposers and fungi that live underearth will not toil, the transformation of death into new life will cease its spinning.

Darkness Lost

The majority of people on most of the planet now live within a constant glow of artificial light, and we are just beginning to understand how this "eternal summer," as scientists call it, affects the inhabitants of earth—both us humans and other beings. Bird migration is among the most studied interruptions of animal life patterns in the face of unnatural light. Under the best of circumstances, migration is fraught—a delicate balance of energy and luck. Journeying birds walk the slenderest equilibrium: enough food to make the next resting place, but not enough to add weight that would make flight too fatiguing; fair winds and weather; fruitful stopovers. All of this is changeable and uncertain. Young seabirds fall into the ocean from exhaustion, warblers die of thirst, hummingbirds of hunger.

Anthropogenic changes have made migration even more perilous. Many birds fly from ancestral memory to forested stopovers and find them covered by concrete, by warehouses, by condominiums; razed by climate crisis–fueled wildfires; transformed into barren monocultures; or simply too denuded to offer sufficient nourishment. Most songbirds migrate at night, and their flights are interrupted by confusion from the constant night-light of urban centers. A recent study suggests that up to a *billion* birds die every year as a result of collision with buildings, a majority of these deaths attributed to artificial light that disrupts circadian rhythms and dissembles visual migration markers. Birds mistake streetlights, office lights, house lights, hotel lights for the glow of the moon and the stars, the play of shadow upon water, or the luminescence of the ocean. There are few places in the entire northeastern US (where the bulk of research has been undertaken) where migrating birds are not confounded by the preternatural glow of a city.

It's not only birds who struggle. We speak in metaphor of being "drawn like a moth to the flame," and we see moths entranced and fluttering about our porchlights. But moths are not *drawn* to the light; they are, like birds, confused by it, their navigational systems scrambled. Gathering around outdoor lamps, moths are eaten by predators, or die from overheating. This interruption of their natural cycle ripples into all of life: we are just now coming to apprehend the importance of moths in pollination; they carry pollen greater distances than bees do, helping to increase diversity and decrease botanical inbreeding.

As far as other creatures go, we have little idea of the extent to which the relentlessness of artificial light affects most of them, but we know the consequences are rampant. Years ago I worked for the Department of Fish and Wildlife in the remote Northwestern Hawaiian Islands, on tiny Tern Island, where four of us were housed in an old Coast Guard LORAN station on just forty acres within a shifting coral atoll. We kept close track of the endangered green sea turtle nests and their projected hatch dates (about sixty days from the time the mother turtle laid her eggs) so we could be sure to turn our few lights off at night, as the hatchlings—tiny volcanoes of perfect miniature sea turtles bubbling out of their dark nests beneath the coral sand—would become disoriented by any light that called them away from the glow of the ocean. I remember gathering the baby turtles in my skirt and running back and forth between the water and the nest as more of them flowed forth, blessing each one as I placed them near the water so they could crawl the last foot themselves into their wild—and most likely brief—new lives. (Very few survive, but those that do can live to be hundreds of years old.) Countless unguarded nests near more populous places bring forth baby sea turtles that are lured toward artificial light and desiccate upon sunrise.*

Trees, too, evolved to flourish within a regular transition from day to night, the rhythms of the sun and moon. Trees and plants measure light on a molecular level, where the length of the day allows them to know the season and follow their natural cycles of photosynthesis, spring budding, and the coloring and eventual shedding of leaves. Anthropogenic light disorders all of these things. Confused by a seemingly extended day length, flowering patterns change, and growth continues beyond the typical time frame, preventing the essential dormancy that allows trees to survive the challenges of winter.

We are only starting to understand the unique ways that animals and trees and other plants apprehend the world around them. Beyond biological survival, we must ponder, too, the keen, inspirited responsiveness of our plant and animal kindred to anthropogenically disrupted patterns of light. What must they feel and know of this? Do they miss, in an animal-tree way, the absoluteness of shrouded darkness that they evolved to inhabit?

---

* Why did we carry the turtle hatchlings to the water? It was part of our work to help this endangered species reach the ocean before they were picked up and tossed like popcorn into the long bills of the waiting frigatebirds (who have plenty of unendangered things to eat).

Human Animals in the Dark

Our human-animal selves, too, require darkness. We have all experienced times in the dark when our beyond-vision senses are heightened: the inhalation of fragrance; the intricacy of earth beneath our feet as in my nighttime moose walk; the songs of the earth folk—trees and rodents and night birds—more bright and layered and loud; our inner sight; our touching of anything at all. Everything becoming more acute. There is a reason we turn off the lights for sex, and only part of it is worry that our abs aren't tight enough for sunlit scrutiny. The organ of our skin is more alert in the darkness as we reach for one another without the distraction of feeding eyes—our skin-pelts seek the grounding exhilaration of another body.

Beyond all of this, new research shows that our physical health may be dramatically worsened by the world of eternal summer we have created with artificial light and ever stable indoor temperatures. In a 2016 study published in *Current Biology*, scientists in the Netherlands showed that "the absence of environmental rhythms leads to severe disruption of a wide variety of health parameters." Mice kept in constant light for several weeks showed increased inflammation, depressed immune systems, muscle loss, and osteoporosis—all of the signs of "frailty" that come with age. With over 75 percent of the global population exposed to light during the night, a deficiency of natural darkness may be making our physical bodies age faster. And although the researchers point out that the studies have not been carried out on humans, there is other evidence to support the direct human response to lack of darkness.

The psychological torture of keeping prisoners in constant bright light is based in neuroscience that shows we become not only psychologically unstable without darkness but also physically debilitated: blood pressure rises; immunity fails; heart attacks, stroke, and cardiovascular disease increase; metabolism becomes unbalanced; generalized pain in the entire musculoskeletal system emerges; the likelihood of death from any cause is dramatically heightened. Dark deprivation in prisons is an overt form of psychological torture and a stealth form of physical torture.

Life at the Speed of Dark

Humans may not be nocturnal, but we are creatures of night as much as of light. Darkness places us upon a disorienting soil, bringing solace and discomfort alternately or concurrently. I write in a time of global uncertainty, within a crisis of ecological collapse and—at the moment—under quarantine during a global pandemic. I write against a backdrop of systemic racism leading to a level of civil unrest and protest unprecedented in my lifetime. Everyone from self-help gurus and Vipassana meditation instructors to my own wise therapist suggest that we learn to be "comfortable in discomfort." But is this the highest counsel? Buddhist ecophilosopher and activist Joanna Macy points out that in much of the spiritual teaching, "grief and anger over current social and ecological conditions are then seen as *attachments* and judged to be less valuable than experiences of tranquility." But this, she says, is a spiritual trap. Instead, "beauty and terror arise within us" simultaneously. It is right that we are disoriented in these times. How could it authentically be otherwise?

Maybe sometimes we don't need to grasp at comfort within difficulty; maybe sometimes we are just uncomfortable. Maybe sometimes we are wholly anxious or despairing. Even in my darkness-walking experiment, though I might not have been actually scared, I was also not calm — my senses were on high alert. In this precarious time, darkness offers the chance to shed our clinging to banal incomplexity. Buddhist professor Stephanie Kaza found this in her own night-walking experience: "Just let the ground support me," she would say to herself. "Walking in the dark night is a way to practice faith, a way to build confidence in the unknown. . . . I learn to practice courage in the vastness of what I can't see."

While light travels, there is no "speed of dark." Darkness becalms us in a constant, receptive awareness. Darkness offers an intelligent stillness that fills and tills our psyche in a manner both difficult and beautiful — this is the *fruitful* darkness Joan Halifax speaks to in her lustrous book by that name. This complicated moment on earth is no time to retreat into the simplistic metaphor of "bringing light." The hope we must maintain, the imagination we must put to use, and the physical health we require all ask of us a more intricate wisdom.

Let us dwell in this darkness as we can. Instead of the luminous as a measure, we can turn with more nuance to the numinous. Seek the dark. Be awake in its ancient cradle. Cozy (or uneasy) in bed with windows wide open to the night sounds. Step into the night, the backyard, the urban sidewalk beneath the same moon that shines upon moose, bears, sleeping falcons, waking owls. Turn off lights, outdoors *and* indoors, during avian migration times—tell everyone. Walk beyond sunset, camp in darkness, sleep outside. We can bring respect to the dark each night, interrupting it as little as possible with light-sound-words-screen. We can listen to the artful messages of our dreams, allow them to affect our day-life. We can protect the night dwellers with knowledge and presence and wonder and love and conservation. Fine soil, clean water, all the way down. Our day actions reverberate into the night. We can make them generous and kind and wise, even as we ourselves are fruitfully unsettled.

In the voice of her seven-year-old self, Opal Whiteley wrote in her diary of the blind girl who was her friend, yet who was afraid of the shadows she could feel with her heightened, nonseeing senses:

Today near eventime I did lead the girl who has no seeing a little way away into the forest, where it was darkness, and shadows were. I led her toward a shadow that was coming our way. It did touch her cheeks with its velvety fingers. And now she too does have likings for shadows. And her fear that was is gone.

STEP INTO THE FRUITFUL DARKNESS

Relate

An Infinity of Animal Intelligences

Walking a forested seaside trail on San Juan Island, my mind finally began to still after a flurry of work on this book in the little cabin where I was hermiting for a few weeks. About half a mile along the path, I was surprised to see the dark cloud of a turkey vulture low on the ground (when not feeding they are usually high in the trees). In my experience, the closer you get to most birds—songbirds, shorebirds—the smaller they seem. Not so with turkey vultures; they are huge animals.

This one was great and round, ruffled with dismay at my presence. As I slowly continued my walk, the vulture lifted and lighted again, just a few feet away. He kept his eyes trained on the inner woods while taking me in, and his reluctance to leave was an easy clue: good, dead food close by. I looked around, saw nothing, and, without even thinking to do it, invoked an animal sense beyond sight—I drank a deep inhale, sniffing the air like a wolf, like a weasel, like a mouse. My sense of smell is paltry in comparison to any of these, but even a sheltered urban vegetarian girl like me could recognize this scent: blood, new and fresh.

The Doe

I've spent enough time observing turkey vultures to know that they will hold their ground; I was sure my investigating would not put this one off—he would simply wait me out. Following the scent, I bushwhacked through salal and sword ferns, uncertain, inhaling again: a little farther and, yes, there in a clearing the size of a cave dwelling was the bleeding body of a small doe. Her swift, light hooves muddied, she was lying on her side as if in sleep, not in the sometimes unnatural pose of death.* The only awkward angle was that of the arrow in her side. The blood I had smelled rivered slowly from her belly—the earth-wild fragrance of life-into-death was so thick, so vital. I stood for a long second, then suddenly overwhelmed with grief and unable to look, turned and scrambled back to the trail.

Within a few days I would not be able to approach this place—the stench of decomposition would be too strong. But today something stopped me from running away and I turned again, rushing back to the doe. I placed my palm upon her withers—lithe, warm. The arrow was an affront; I could not leave it. Pulling it out took two hands, all my strength. It must have lodged in a rib—after my initial struggle it came out quickly; more blood, more smell of blood, and I started to cry. "Fucking idiot hunter, fuck fuck fuck." (This was University of Washington land where hunting is illegal; and this doe was not, I

---

* Deer do not actually lie on their sides to sleep but, instead, with their legs tucked beneath them. This deer's position evoked that of a deep-sleeping person.

later learned, the first black-tailed deer to die after being shot by a hunter cowardly enough to care more about not getting caught than about letting a sister animal suffer. I learned, too, that this was not a clean shot as far as bow hunting goes; it was unskilled, and should never have been attempted, even by one who pursues hunting as sport.) The anger washed over me fast, then left me in stillness with the deer. Other than this meandering blood, she looked so perfect, so near-to-life.

I sat cross-legged at the doe's head and lifted it—heavy—onto my calf, stroked her cheek, long lashes rimming her wide, dark, open eyes. Soft ears, the softest. "You have done well, little one." I heard the words from my lips, unawares. "Roam now in peace and beauty."

Her fur was every color—fawn, brown, black, blond, ochre, even white. Lighter hairs seemed to congeal into the hint of a star on her forehead. Now the vulture flew to a branch just over my head in this small forest tomb, then two more—not at all silent in their arrival. They fluffed their feathers. It was time for me to go. I kissed the deer's star-head and raised a hand in silent homage to the waiting vultures.

The Orca

In the 1960s and 1970s, forty-seven Salish Sea orcas were captured for display in aquariums across the country. In one horrific day at Whidbey Island's Penn Cove, forty whales were violently separated from their family groups as the public looked on. Some of the watchers wept, some cheered the captors. The orcas screamed, bled. Mothers rushed to save their young; no orcas abandoned the others but set their bodies between the nets and their kindred. The captured were torn away that day, bloody, terrified, leaving their tightly woven pod-community in fear, confusion, and grief. Many died.

Today, the only survivor of this brutality is an orca who was named Lolita by the industry that holds her captive. She is kept in a sixty-foot tank at Miami's Seaquarium (she herself is twenty feet long), where people clap and laugh as she swims in heart-breaking circles, separated from a community of orcas known for swimming always side by side. The Lummi Nation has named her Tokitae, "lovely day, pretty colors" in Coast Salish, and have proclaimed it their sacred obligation to return her to her ancestral waters; the Lummi have joined with Orca Net-work, and together they have researched a detailed plan to bring Tokitae back safely. Her pod of endangered southern resident killer whales, the L-pod, still respond to recordings of her voice, and Tokitae, after forty years, still sings the unique language and dialect of her extended family. She remains imprisoned.

Looking back to the moment when I ceased running from the sadness and horror that the freshly dead deer inspired in me and turned back to her, I remember the Lummi's teaching about Tokitae. This orca is a sister, a relation, to whom we are obli-gated in the sacred belonging of kinship. When the Lummi speak of her, and of all orcas, there is no compulsion to explain these terms — sister, brother, relation, family — academically or ratio-nally. This knowledge is spoken with the authority of the Haida teacher who earlier put us all in one metaphorical canoe: *We know this, and it is true.* It is *this* that turned me back to the dead deer, as it would have for so many of us in such a situation. Not the sense that removing the arrow was the humane thing to do (reflecting well upon my own humanity or my species as a whole), not out of my volatile passing anger, but out of a rooted sense of absolute relatedness and the innate, unbidden, whole-hearted love this ordains. I turned back to my deer-sister because it was right to honor her life and death. The moments after her passing would not unfold in the presence of a human horrified and afraid, but of a sister standing, then kneeling, in love.

Beyond Anthropomorphism

In 1637, when French philosopher René Descartes wrote "I think therefore I am," he relegated consciousness to the processes of human thought, creating a dichotomy between matter and mind. Spirit and soul were woven into Descartes's definition of thought, and his ideas found a ready foothold in the religionized philosophy of his time; the chasm between spirit and matter became solidified. Humans alone possessed consciousness, humans alone were ensouled.

With Descartes's scientific method paving the way, Men of Knowledge could dissect a dog, strapping her down and cutting her open with impunity, dismissing her wails of pain as nothing more than, in Descartes's words, "the squeak of a door hinge." Only in the last few decades has biological science begun to recover from this influence and speak of animal consciousness, thought, intelligence, feeling, and emotion without banishing such things to the academic dustbin of soft science. Until very recently, any scientific reference that might be perceived as a nod to animal consciousness was enclosed by diminishing quotation marks: *The monkey appeared "happy,"* thus distancing the writer from the cardinal sin of anthropomorphism, the attribution to animals of emotions or traits thought to be distinctively human.

Yet many naturalists after Descartes who spent time in deep study of various life-forms bridged this gap. Charles Darwin thought it so obvious that the continuity between all creatures and humans included not just morphology but consciousness that he never even bothered to defend the idea. It was, for him, a given (and Darwin's own "tree of life" was always more of a shrubbery — with humans in the mingle of mosses, fish, and elephants, rather than perched precariously atop).* Darwin often referred to animals as being happy, calm, peaceful, or agitated (not "agitated").

In 2012, nearly 150 years after Darwin, an international consortium of scientists signed the Cambridge Declaration on Consciousness, proclaiming that all animals, from birds to apes to coyotes to fish to octopuses, possess consciousness worthy of ethical consideration. It was a welcome step, to be sure, but also, one cannot help thinking, a bit late. In his 2007 book, *The Emotional Lives of Animals,* evolutionary biologist Marc Beckoff

---

* That happened later, when eugenicist Ernst Haeckel and others combined Darwin's natural selection and Aristotle's early metaphysics of ascending, immutable characteristics in organisms to create a tree topped, in capital letters, with MAN.

put it so simply: "Emotions are the gifts of our ancestors. We have them and so do animals. We must never forget this."

Observations labeled anthropomorphic are very often not a sentimental or unstudied imposition of human mind states upon an animal but simply a *recognition* of emotions and reactions to the ups and downs of biological existence that we all share. As animals ourselves, we know when we see contentment, relaxation, calm, fear, pain in many other animals. That said, describing animal states in human language can go awry in two related ways: we can diminish the soaring uniqueness of other species and individuals when we assume they are intelligent only to the extent that we can perceive in them humanlike ways of knowing and feeling; and we can fail to recognize or even imagine the breadth of unique animal intelligences because they lie outside of human measures and ways of being.

Certainly it is possible to overattribute or misattribute various mind states to other animals; discussion of other-than-human animal consciousness requires our highest intelligence and our deepest discernment. But I am ready to abandon use of the word *anthropomorphism* altogether. It is time to acknowledge animal consciousness—both the continuities that we share and recognize, and the mysteries that we may never comprehend—without hindrance from the strictures of such a term. It is time to speak of animals in the way that is *most true*—open to all that we can know and respectful of all that we cannot—without the constant need to defend ourselves against outworn language or outdated science and philosophy.

A Multiplicity of Intelligences

A starling lives in my house. Mozart also lived with a starling, and I wrote extensively about the cross-century relationship between these two birds (as well as humans and animals) in my book *Mozart's Starling*. During my research, I'd learned of a starling nest at a nearby park that was going to be swept away by city officials even though the nestlings had hatched, so I thief-rescued one of the four-day-old young. Of course, under the Migratory Bird Treaty Act (and common ecological sense) it is entirely illegal to disturb the nests, eggs, or nestlings of nearly every bird in North America. As an introduced and invasive species, starlings are different—we are encouraged by fish and wildlife departments to keep them from nesting, destroy their nests, and even kill young and adult birds. I agree with finding ways to discourage the nesting of invasive birds but am against harming them once they are born, which does not help control their populations in any case. We have created the conditions in which starlings flourish; ecologically minded landscape solutions are required to make a lasting difference with this lawn-loving bird.*

Along with parrots and corvids, starlings are among the most gifted mimics in the avian world. Carmen mimics human lan-

---

* Still, while we are allowed to kill, torture, or maim starlings with legal impunity, it is decidedly *not* legal to lovingly raise one in your living room. Though I have published a book about Carmen's escapades in relation to my household and to the great composer, and spoken of her in lectures around the world, the starling police have yet to appear.

guage phrases, especially those we most often use while interacting with her (starlings love interaction): *Hi sweetie, Hi honey, C'mere.* She mimics household sounds: the coffee grinder, the creak in our old oak floor, and the *psst-psst-psst* sound we use to call the cat. She mimics the song of the Bewick's wren that nests in the camellia outside her window.

It is believed that starlings mimic one another and sounds from the world as a way of sharing information and bonding socially. The creation of a varied and aesthetically impressive personal repertoire helps males attract a mate. But living with a starling all day every day, I learned something rarely recognized in the scientific literature: Carmen not only responds to the sequences of sounds she hears, but *anticipates* them — participating actively and aurally in the daily round of our home life. Early every morning, I walk down the dark stairs and Carmen whispers in a hushed morning voice, *Hi Carmen,* the first thing I would normally say to her. Then, when Delilah, the hungry tuxedo cat, enters the kitchen, Carmen calls in a loud, hungry cat-voice, *Meow!* When she hears the tinkle of coffee beans in the jar, but before I start the grinder, she calls a coarse *Whiiiiir!!* (Not her most musical vocalization.) When I open the door of the microwave, but before I set the timer, she sings *Beep! Beep beep!* Pitch perfect. It dawned eventually on my slow human brain that it is not just this precocious young starling but *all* starlings who have such astonishing aural attunement to the world and all that passes within it. And this is just one animal with one way of being, a way that I just happened to become aware of, living in uncommon intimacy with a single wild bird.

;;;;;;;;;;;;;;;;;;;;;;;;;;;;;;;;;; (This last is courtesy of Delilah, walking across my MacBook, and since this is a story that appreciates intraspecies offerings, I thought I'd let her make her presence known—semicolons are a suitably catlike, noncommittal form of punctuation.)

The starling's way of being in the world is unique—as is that of all beings. As a species, turkey vultures like the ones I encountered with the deer in the woods vocalize little, but their brains hold more space for sense of smell than any other bird, and they are called to their flesh-meals through aromas that rise from earth to air. But this sense doesn't simply turn off in vultures after a good meal. My own olfactory sense is trifling next to a vulture's. How must it be to live guided by flight and fragrance? What manner of intelligence forms within a life shaped and molded by these things? Or the whisker-based night-seeing of rodents? Or the skin-based knowing of earthworms? Or the beyond-human hearing of bats? Or the geometrical pattern recognition of bees in the flowers they see, and the visual wavelengths that humans are blind to but that guide bee-lives?

Each morning I hear Carmen speak in her own responsive voice, a graced reminder that I walk always within an infinity of animal intelligences, some known, some never suspected, some unimaginable to me in my limited human comprehension. These unique ways of knowing encircle our everyday lives in shimmering complexity.

Coming into Animal Presence

Ever since humans have lived upon earth, we have made our homes and conducted our movements in proximity to other animals. The more prominent our enclosed modern dwellings, encapsulated modes of transportation, indoor workplaces, and ever-present technology become in everyday life, the more we are separated from the presence of other animals who have always been a part of human life-making. The beloved domestic dogs and cats who share our homes are a delight, but no substitute for time alert to the vivid intricacy of wild visitations and interactions.

We are experiencing now an isolation named *species loneliness* by Michael Vincent McGinnis in a 1993 paper for *Environmental Ethics*. In his book *Our Wild Calling,* Richard Louv describes this modern human condition as "a desperate hunger for connection with other life. . . . All of us are meant to live in a larger community, an extended family of other species." Without this, a number of pathologies grow within us and "the family of humans loses comfort, companionship, and perhaps even the sense of higher power, however one defines it." Animals, too, have evolved with humans among them — and this distanced relationship in which we currently live may be an incalculable, unknowable loss to them as well.

Being in the presence of wild creatures, even those most of us experience in the simplified close-to-home wild, is not always comfortable, comforting, or the least bit straightforward. In the presence of any animal at all, we might experience wonder, awe, fear (the relative of awe), preternatural calm, unfamiliar joy, the impulse to draw nearer, the impulse to run — in succession or all at the same time. I keep thinking of those rattlers, that moose, the shark shadows that have swept beneath me while snorkeling, the bears I have passed on the trail, the baboons who walked through my volunteer hut on the coast of Kenya, stealing the fruit and hissing at me through knife-sharp teeth.

Recovery from species loneliness will mean walking with a courageous new tolerance for complexity and discomfort, an allowing of difference, a delight in sameness, an openness to wonder, being alert instead of being afraid.

This is a tangled empathy. Ecopsychologist Patricia H. Hasbach created an unusual list of engagement patterns shared by humans and many other animals, not just those with eyes and fur; Louv collected and added to these in *Our Wild Calling*. Alongside many animals, we: recognize others and know that we are being recognized in return; have a natural and mutual curiosity; play with our own species and often with others; share empathy; and use "intonement," a humming or chanting of notes for a variety of purposes. In addition to all of this, we are able to "cross the threshold," as Hasbach says, to enter the "psychological or spiritual space" that hangs between two animals. This grounds another path to empathy on Dr. Hasbach's list, perhaps the most inscrutable to modern culture: "Becoming the animal."

There is only one certain way to find intimacy with the wild ones among us. We can, of course, plod along the sidewalk or run up a trail or glance up from our desk and, in a chance instant, spot a gull or a warbler or even a coyote. But if we want to make regular observations and deepen our understanding, then we must be present, and we must be still. And then more still.

Jon Young, cofounder of the Wilderness Awareness School, focuses on what he calls the "sit spot" as a core routine in nature connection, the root of all others. Choose a spot outdoors—as close to food, shelter, and water for wildlife as possible. In urban places this could be a bench in a sidewalk garden, a patio overlooking a tree, the edge of a vacant lot. We all discover our wild corners. The school's wonderful central text, *Coyote's Guide to Connecting with Nature,* counsels: "Find one place in your natural world that you visit all the time and get to know it as your best friend. Let this be a place where you learn to sit still—alone, often, and quietly.... This will become your place of intimate connection with nature." Young invites us to look into our childhood and think about whether there was a place that we went to regularly, just to sit, maybe under a tree. I had Frog Church. And for years as an adult I have had two sacred places—one in my backyard and one in the wooded park down the road. Young considers this the "magic pill"—the practice that grounds all others.

Most of the creatures that coinhabit our landscapes have a sense of where the humans wander, and they often avoid these edges, nestling toward the inner woods or other protected places. Much of the animal behavior we regularly observe is the stilted, watchful behavior that occurs in the presence of us large bipedal mammalian predators. Animals stop feeding when we turn up; they stop moving; the winged ones fly to a comfortable height or distance. Yet when we are still for a long time, the interruption caused by our human presence ebbs, and the animals among us return to their natural ways of being in the world. We are privileged to stand witness. The still spot is an enchanted parenthesis between worlds—the everyday world of distraction and the ground of an empathic, wilder life.

Come to your still place every day, at all times of day, in all seasons, in all weather. Just have fifteen minutes? That's OK, but longer is better (and much longer is much better). Let the earth settle beneath you and any creatures return to their tiny detailed life patterns. Sit, breathe, watch. Having left your phone behind, it will be easy to resist the overwhelming impulse to tweet or Insta the moment. Over time, the nuances of daily changes will be revealed. Ingrained aversions to a bit of dampness or direct sun or the tickle of an insect on your skin or an unexpected rainfall or even a small spider in your hair will evolve into intimate conviviality. Soon, when a bird arrives, even an individual of a common species, you will know whether it is a "regular" or someone new to the place. And if you are faithful to this practice over time, many of the birds will eventually respond with the wild grace of recognizing *you*, remaining settled and calm on your approach, knowing that you are the Being Who Sits Still and Is Always Quiet.

Shapeshifting

The wonderful British mythologist, author, and artist Terri Windling writes:

Folk tales from around the world tell us that the animals communicate with each other in a language unknown to men and women—or else in a language that used to be known to us, but is now lost. The stories also tell of human beings who understand the speech of animals. Some are born with this ability, while others obtain it through trick-ery, or magic, or as a gift from the animals themselves, a reward for a great act of kindness.

What if this great act of kindness is not something as out-wardly heroic as pulling the thorn from a lion's paw, as in Aesop's fable, but something just as exceptional in our time? What if it is simple stillness, presence, a rare allowing of animals to reveal *themselves*?

Here we come back around to that mysterious and somewhat unnerving phrase from Hasbach's list: "becoming the animal." We are, of course, already an animal. Yet through cultivated qui-etness, through observation, through art, through story, through myth, through poetry, through intuition—and without any need to appropriate rites played out with depth and care in shamanic cultures—we can go further and greet the slender times that we fall into prerational interbeing with another animal.

This is shapeshifting. It does not lie outside of the everyday magic of earthen existence, and is the reason that shapeshifting tales, such that of as the Gaelic selkie, are so resonant. The seal-woman sheds her skin to dance upon a stone with her selkie sis-ters, when her pelt is stolen by a hunter. She marries the hunter, lives for a time without her seal pelt on land, but begins to wither. The selkie is made whole only when she finds her authen-tic animal skin, her soulskin, slips back into it, and then into the wild waters. In *Foxfire, Wolfskin, and Other Stories of Shape-shifting Women,* mythologist Sharon Blackie reimagines a simi-lar Croatian tale, featuring a wolf-woman instead of a selkie. In this version, the woman eats the husband who stole and hid her wolfskin! Dr. Blackie notes that these tales most often end with the rewilded woman simply leaving the skin-thieving husband to contemplate the error of his ways. But Blackie is mischievous: "I've always preferred stories which come with consequences," she says.

In her essay "Forests of the Mind," for *Aeon*, Jay Griffiths emphasizes that shapeshifting is possible for all of us as "part of the repertoire of the human mind, cousin to mimesis, empathy, and Keats's 'negative capability,' known to poets and healers since the beginning of time." Griffiths notes that we are speaking not of literal fact but of a "slanted, metaphoric truth." She invokes Rilke's "Second Elegy," where "we, when moved by deep feeling, evaporate; we / breathe ourselves out and away." Shapeshifting is a kind of "transgressive experience," Griffiths writes, a "crossing over," a "self-forgetting and an identification with something beyond." In this state, we become receptive to wild ways of being beyond our semipermeable skins, and more knowing of the ones already within us. It is a welcome bewildering, a liberating new wisdom. A redemption.

Lifting Up

I was talking this week with my dear friend Lynda Mapes, a journalist for the *Seattle Times* who recently won the American Academy of Sciences award for her series "Hostile Waters," about the endangered southern resident killer whales of Puget Sound and the wider Salish Sea. It was Lynda's writing that brought to the wider world the story of Tahlequah, the young orca whose daughter died soon after she was born. In an act of grief, loss, and possibly resistance, Tahlequah carried her newborn for a thousand miles and seventeen days before finally letting her go. Lynda believes that Tahlequah did not give up carrying her baby but that after more than two weeks the little body simply fell apart. Tahlequah was accompanied in this pilgrimage of grief by her community, the J-pod.

As I write these words, we have just learned that Tahlequah has given birth to a new baby, a beautiful son named Phoenix. He seems healthy and precocious, but, knowing the threats that this population faces, and after the heartbreak of Tahlequah's last birth, we are all holding our breath. In her work Lynda has covered tribes and the environment for decades and has followed the region's orca population for twenty years. She has friends among the Lummi, the Orca Network, NOAA, other scientists, activists—all of the orca people. She is positioned as well as anyone to speculate on the question we all share: what are the chances that Tahlequah's new baby will live? Births among southern residents in recent years have been few and mostly unsuccessful—miscarriages, stillbirths, early deaths. Chinook salmon, the resident orcas' primary food source, are increasingly scarce. In this time of pandemic, attention and resources have turned from the land, waters, and creatures to basic human survival. Yet as everyday shapeshifters, we interweave our thoughts with the orca's body: what will happen to Tahlequah's baby?

Lynda was quiet for a while, then switched up the question. In her interactions with the Lummi People, she has learned to stand outside of the calculus we all want to grasp on to: the balance of resources, the health of the waters, past outcomes, future projections. Instead, we are asked to hold Tahlequah and her baby up. Our dear sister, our small brother. To lift them up with love in our hearts and our spirits. And that's it. This lifting and the actions that follow from it: *this* is our sacred obligation to all our relations.

LIFT UP OUR ANIMAL KINDRED

Speak

The Abracadabra of Earth

While on faculty at a writer's conference recently, I attended a talk by another teacher who claimed that the magical incantation *abracadabra,* made familiar by cartoons and magician parodies, means, at root, "what is spoken is what becomes." While this is certainly a happy thought for writers looking to justify the significance of our word-work, it seemed too good to be true. Yet when I dug into etymological sources and consulted classical language scholars, I found it to be credible. *Abracadabra* is from the Hebrew-Aramaic lineage, murky but traceable. What we say is what we bring into being. The way we speak shapes our perceptions, our actions, and ultimately the outcomes we seek.

The association of language with enchantment is an ancient one. "Chant" is from the Old Northern French *cant*—a sacred singing that (with *en-*) allows transformation. The word *grammar* that frames our language use inspired the word *glamour* to describe occult power; for the illiterate in medieval times, the language of those who could read and write was magic indeed.

Language lives and evolves in many of the same ways that organisms do: by mutation (when a random genetic change catches on); by geographical isolation (as when animals or plant seeds cross water, do not interact with their mainland counterparts for some generations, and become biologically distinct and endemic to their place); and by practical usefulness (in a neo-Lamarckian fashion). Tracing the etymological evolution of a word can lead to understanding its meaning with surprising new depth. A word is as alive as a bird.

True Naming

Ursula Le Guin recognized the significance of "true naming" and the mystical strength of language. This thread runs like mycelia through all of her work but most overtly in the Earthsea world, where all people and all things have an everyday name and a true name — a soul name. Use of the latter can enhance the delicate balance of existence or thwart that balance. A person shares their true name in sacred confidence, only with someone they trust with their life. The wizard Sparrowhawk, whose true name is Ged, avows:

It is no secret. All power is one in source and end, I think. Years and distances, stars and candles, water and wind and wizardry, the craft in a man's hand and the wisdom in a tree's root: they all arise together. My name, and yours, and the true name of the sun, or a spring of water, or an unborn child, all are syllables of the great word that is slowly spoken by the shining of the stars. There is no other power.

Words and language open possibilities, but also limit them. When we use words in certain ways, we lay down the roots and trails of how these words will shape the world, and how the ideas they influence will—and *can*—be used. When we choose one word over another, we offer more than a simple building block for a sentence. We offer a constellation, a worldview, a story, an action plan.

My dad is a stonemason, and well versed in moving heavy things like rocks and bricks around in wheelbarrows. Ever since I was little and pretending to help in the yard, he has tried to teach me a simple maxim: Always point your wheelbarrow in the direction you want it to go *before* you fill it up. But damned if every time I am in the garden I turn to push my beloved little red wheelbarrow to the compost bin and it isn't facing the wrong fence. Our words (like our wheelbarrows and our magic wands) must point in the way we want our precious life-effort to take us.

I hear over and over from thoughtful friends and activists that if we don't couch our dialogue about nature in terms of utility and economic value—whether the profit be derived from eco-tourism, or carbon sequestration, or potential for pharmaceutical discovery, or growing board feet of lumber, or even human health—then wild places and creatures will not be saved. As activists, we have tried using the utilitarian language of economics in order to protect nature for more than a hundred years. We have tried this since the eighties in specific efforts to protect Pacific Northwest old-growth forests. It has not worked. As long as we frame a worldview with language that refers to the wild as a commodity, it will be treated as one.

It is likewise damaging to invoke technology-based metaphors to explain nature: the brain a computer, the earth a spaceship, the rooted and fungal soil beneath our feet a kind of internet. Such mechanistic phrasing unwittingly invites us to see the natural world as other-than-alive and reparable by human skill in ways that it simply is not. If we are seeking a relationship within the earthen community that is meaningful, genuine, and impactful, then the words we use to describe that relationship, and the beings in its purview, must be chosen with intention, with specificity, with intelligence, and with love.

Language is a central foundation of a new story that proclaims our deep interflow with the wild earth and allows a sense of wonder and kinship beyond mere utility, even as we feed, clothe, and house ourselves with the stuff of the natural world. The tendrils of the words we use now will reach far — into the psyche of the mind and the future of the earth. They need to be right.

Excavating True Names

In his book *Landmarks,* Robert Macfarlane speaks to the need for "place-particularizing" language. When we dumb down the language of nature, our understanding, relationship, longing, and love for the earth are likewise imperiled. By contrast, when we call things by their specific names, we locate them in place, in lineage, in memory. We enkindle kithship. When I am walking in the world and see a thing and know its name, I experience a rush of intimacy. I *know* thee, redcedar, sword fern, varied thrush, just as I would a friend's name, a neighbor's, a sister's.

Yet just because a natural organism or land formation *has* a specific name doesn't mean that name must be accepted as a good one. Many of the names given to creatures and landscapes do not ring true. The vast majority of official species names in North America were made up by white, English-speaking, male scientists or explorers, claiming to have discovered organisms well known by people who had lived among these same organisms for thousands of years and had already worthily named them.

In the US, there are nearly a hundred birds with American Ornithological Society–ordained names that recognize white male explorers and ornithologists: Swainson's thrush, Steller's jay, Cooper's hawk, Bewick's wren, Lewis's woodpecker, Clark's nutcracker, Cassin's auklet, Bullock's oriole, Cook's petrel. On and on. Many of these longstanding names are deeply fraught. John Kirk Townsend (Townsend's solitaire, Townsend's warbler) stole remains from the graves of Indigenous people. John P. McCown was a Confederate officer and a defender of slavery, and he battled Indigenous tribes. Critics only recently succeeded in convincing the AOS to rename McCown's longspur (now the thick-billed longspur).

Is that true? As soon as I wrote it down I second-guessed myself. Surely someone has done this research, but I became obsessive about answering this question for myself, so I pulled out my field guides and historical sources and emerged two days and twelve hours of study later armed with the knowledge that it is true indeed. The few birds named for women still represent a highly colonized language: there is a duck previously known as oldsquaw—an offensive, racist, misogynist name long in use and finally changed in 2000 to the descriptive long-tailed duck. Anna's hummingbird was named for the Duchess of Rivoli, a patron of exploration, whom artist and hunter John James Audubon (owner of enslaved people, and collector of Mexican soldiers' skulls from Texas battlefields) reportedly found to be "beautiful and charming." Virginia's warbler was named by naturalist and first US Fish Commissioner Spencer Fullerton Baird (of Baird's sandpiper and Baird's sparrow), for the wife of his friend, a Confederate soldier. The same Baird named a bird after his own teenage daughter: Lucy's warbler. The Blackburnian warbler is named for a specimen of the bird sent from New York to London for identification by a woman known only as Mrs. Blackburn. That's it. That's all of them. Women who did deep work in mycology and botany in Victorian times (both in the UK and the US) were sometimes acknowledged in species names, but these names were conferred officially by men in scientific societies that were closed to women.*

In the early 1700s, Carl Linnaeus devised a system for naming species, and he made it his life's work to engineer a vast *Systema*

---

* In the 1829 volume *Familiar Lectures on Botany,* Almira Hart Lincoln reinforced the popular view that the study of botany and fungus was "peculiarly adapted to females; the objects of its investigation are beautiful and delicate; its pursuits, leading to exercise in the open air, are conducive to health and cheerfulness." Study of bloody things, like animals, was deemed less appropriate for women.

Naturae based on this naming; his Latinized genus-species binomial nomenclature is still in use. Scientific names are created and maintained in part to indicate taxonomic relationship (we can tell by their names that the spotted owl, *Strix occidentalis,* and the barred owl, *Strix varia,* are in the same genus and thus closely related) but also to eliminate confusion. In the parlance of common names, a "blackbird" in the UK is a "robin" in the US (both are thrushes). But a "robin" in the UK is from a different family altogether (not a thrush at all). Scientific naming allows researchers across the globe the essential ability to confer with confidence in any language.

North American common bird names have been codified by the AOS—persnickety right down to capitalization and hyphenation, making common names in North America as lexiconic as scientific names. Students of botany are subject to no such regulation, and so a single native plant—say, the osoberry that grows wild in my home ecosystem—might also be called Indian plum, Cascade plum, chokecherry, or Osmaronia. The subalpine mariposa lily is also referred to as mountain cat's ear or sego lily (from *seego,* the Shoshone word for "edible bulb"). Is this confusing? Certainly. But in such common names for species, we find the relationship of particular folkways, local languages, places, stories, values, medicine, craft, cuisine, or a combination of these things. When there is only one official common name (no matter how useful), the uniqueness of relationship to place— kithship—can be lost.

I am all for knowing both scientific and common names; I read field guides like they are sacred texts, and feel an uncomfortable, neurotic lostness when I am in a new place, see a lizard, and can't name the species. Yet scientific naming can be a way of asserting human intention and control over nature; the rightness of these names will always be worth questioning, and they can never be the final word. Linnaeus? Philosopher Susan Griffin reminds us that later in life he suffered a massive stroke and his memory began to fail. "Gradually he no longer knew *Systema Naturae*, and after all this, in his last years, he forgot even his own name."

I've just finished composing a small piece for inclusion in a forthcoming guide to the Cascadian bioregion. Instead of the linear, taxonomic order of species in most guidebooks (where all owls appear on the same page, though one species may be found only on tundra, and another in old-growth forest, with no ecological relationship between them), this guide is inspired by forms of Indigenous taxonomy that recognize kinship clusters — redcedars, sword ferns, varied thrushes, and spotted owls are grouped together, for example, standing in community within their ecosystem and outside of externally imposed schemata. My piece is about the chestnut-backed chickadees; they are a bit like black-capped chickadees who have been dipped in a cedar forest, made over with patches of umber feathers and a certain quietness befitting their conifer-based life. Instead of bearing standard taxonomic names, many beings and entities in this new guidebook will be named by poets, artists, naturalists, Indigenous people, and lifelong residents of Cascadia who have formed kithship with the land.

As we seek to question and deepen the current official naming process, it is essential to excavate local names, and learn and honor Indigenous names. It is worth creating community names, poetic names, and personal names for the creatures and land-marks in our sphere — the names they reveal to us in their mor-phology, behavior, presence upon the land, relationships among one another, and visitations to our psyches.

Speaking of Others

Many animals are easy to distinguish by sex. Males tend to be more showy — ready to win a mate with their shining beauty or to fend off territorial invaders with an impressive display of color. Male lions have manes, elk have antlers, bulls have horns, birds have bright breeding plumage. This we know. But many animals are not easy to distinguish by sex: crows, ravens, star-lings, coyotes, raccoons, squirrels, opossums, bears, cougars. While a seasoned watcher can tell most of these apart by behav-ior or contextual cues or subtle morphological differences, in general the fur or plumage of males and females is indistinguish-able.

Many such animals are common in urban ecosystems, and I see them often while in the company of other observers. I've noticed that if I refer to a crow as "he," no one says anything. If I refer to the same bird as "she," people ask, always, "How can you tell that's a female?" The gender-neutral language we have adopted in our literary and colloquial discourse in recent decades has not meaningfully changed the terms in which we discuss nonhuman animals.

In my own writing, when I don't know the sex of an animal, I will refer to it simply as "he" or "she" in order avoid "it" (and the cumbersome "she/he") as much as possible. An "it" can be an object—nonliving and lacking consciousness. Allowing animals (as well as trees, plants, and other beings, some of whom are neither male nor female but both at once) to become *it*s in both our everyday and academic language creates a barrier to empathy, and an invitation to exploit both the creatures and the land that supports them, even if the trees and animals themselves don't know or care what we call them.

But what to do? Randomly assigning gender in equal proportions may be a step up from labeling everything as male or "it," but there remains a pronoun problem in our speaking and writing as we attempt to create a more authentic language of animacy. The current move toward gender-nonbinary language among humans may facilitate a rethinking of pronouns for other-than-human beings as well.

Inspired by her own Potawatomi lineage, Robin Wall Kimmerer sought the wisdom of tribal elder and language guide Stewart King as she explored alternatives to "it," and wrote about the experience in her essay "Speaking of Nature," published in *Orion*. King offered her *aakibmaadiziiwin*, which he translated as "a being of the earth." And while she felt grateful to discover that such a word exists, she realized that the long phrase would not slip easily into popular English usage. She experimented with part of the word—the beginning, *aaki*, means "land"—and shortened it further to *ki*, signifying "a being of the living earth." For Kimmerer, *ki* substitutes directly for *he* or *she*. Of the crow outside my window, I would say, "*Ki* is so raucous this morning!" Or of the chickadee, "I must put out some sunflower for *ki*." *Ki* prowls the morning neighborhood on soft coyote-foot, *ki* runs the river with shining salmon-skin. Plant-beings, too, and forces like the wind become inspirited with this small word, all alive, says Kimmerer, "in our language as in the world."

Such a word might invite a new way of relating to nature, or perhaps invoke an ancient one. But when Kimmerer tried *ki* out with her students, she found some resistance. It is hard enough to incorporate a new word, let alone one that encompasses a transformative worldview. I am almost positive that if I asked my editor whether I could substitute every living being's pronoun with *ki* in this book, she would scowl. And yet, whether it is *ki* or some other word, we are being pressed on this changing earth toward an expanded intimacy with all beings. As we ponder and play with this idea of *ki* and evolve a rooted language of interbeing, we can, at least, commit to leaving *it* behind.

Behold

The voices that influence our lives and work do not always speak in human language. When I am uncertain what to write, I go to the foot of a bigleaf maple or a grandmother redcedar, sit upon her roots, lean against her wide and wizened trunk, sink into the earth, and whisper, "What do you want me to say?" Then I listen. And what I realize, more than anything, is that instead of speaking in words, we are often asked to dwell in silence, where there is no imposed story, just the earth and us there, too, listening beyond language to one another.

Though the mystical revelations that came to Julian of Norwich are most often described as visions or showings, Carol Lee Flinders points out in *Enduring Grace* that Julian herself would often call them "touchings"—things she experienced firsthand, in "forms of understanding that are not merely cerebral at all." There are elements of the sacred earth that flourish beyond our rational thinking and naming.

Sometimes the withholding of human language may be the best path toward apprehending other forms of communication. Cultures across time have known this. In *The Spell of the Sensuous,* David Abram writes:

The belief that speech is purely human property was entirely alien to those oral communities that first evolved our various ways of speaking, and by holding such a belief today we may well be inhibiting the spontaneous activity of language. By denying that birds and other animals have their own styles of speech, by insisting that the river has no real voice and the ground itself is mute, we stifle our direct experience. We cut ourselves off from the deep meaning in many of our words, severing our language from that which supports and sustains it. We then wonder why we are so often unable to communicate, even among ourselves.

One day I was walking in the woods with my friend Trileigh—a kindred spirit and adept naturalist. She recently achieved emerita status after a fine career as a professor of environmental studies at Seattle University, and we wondered aloud together about next life steps, both of us being at something of a crossroads—she newly retired, and I newly empty-nested, my daughter fledged from home to college. Trileigh has many irons in the fire—she's a Renaissance woman with a passion and talent for photography, art, and writing. There is no worry of her withering away in boredom. Still, there hangs the ominous question of Life Purpose. "I wonder whether part of my work at this point might be to simply witness," Trileigh said, "but I feel I should really be *doing* something." We recalled the words of Mary Oliver, who composed famous instructions for life that include the admonition to *Pay attention, be astonished,* and *tell about it.*

In these lines we are called to witness the earth through personal experience, and—a dimension of the meaning of witness—to testify based on that experience.

These are good instructions. But I wonder: what if we leave off the last part of Oliver's instruction sometimes? To witness can simply mean to be present—no testifying involved. Why not *just* witness—see a thing because we are there to see it, knowing that our presence is a privilege for ourselves and a quiet offering to that being witnessed. Why not allow astonishment to visit as it will, then walk into the world changed but perhaps silent—without thinking up words and ways to "tell about it," or imposing a narrative upon our encounter at all?

This is *beholding.* For spiritual traditions all over the world, such a stance—that of contemplative witness—is in itself prayer, art, and activism. Sometimes, letting go of the urge to name and tell—to embroider our own story upon a being or experience, however unwittingly—allows us to see deeper into the beheld while cultivating a unique and poetic wholeness within ourselves, like that possessed by an owl or a stone.

Try these things: Keep field guides everywhere. And topographical maps. Read them like novels, like holy texts, like poems. Learn the names of new-to-you wild beings or landmarks in your home region, then create your own living names for these same things. Respect Indigenous names. Listen for the earth to whisper a new name for yourself, and tell it to everyone or to no one. When you are at a loss, put your ear to the forest floor, or the bark of a tree, or tilt it toward the clouds. See what wordless language points you along your path.

SPEAK IN TRUTH

Grow

Trees and the Rings of Belonging

The park where I take my near-daily woodland rambles was designed by John Charles Olmsted, stepson of Frederick Law Olmsted (and by many accounts the most gifted landscape architect in the famed firm). Olmsted appreciated the native forests of the Northwest, but had no issue with keeping farm-relic fruit trees and escaped garden cotoneasters in the park while clearing space for species not found in the region — coast and sequoia redwoods, incense and yellow cedars, American beeches from the eastern US.

Overall, the park is a rare urban forest, with native old-growth trees and the classic Pacific Northwest evergreen trifecta of Douglas fir, western redcedar, and western hemlock everywhere in predominance. These intertwine with their native forest partners — moss-covered bigleaf maples with the epiphytic licorice ferns that blanket their branches; rangy red-peeling Pacific madrones; and a classic understory of vine maple, elderberry, osoberry, sword fern, salal, trillium, salmonberry. (Naturalist's trick: when the salmonberry flowers, the migratory rufous hummingbirds arrive; when the flowers ripen into berries, the long-awaited Swainson's thrushes, with their magical faerie songs, appear.) Modern wanderers might wish for an all-native park, but Olmsted had an artistic eye and the nonnatives do not much intrude. We are blessed by this place.

The Grieving Trees

On a recent day in the park, I greeted and passed the enormous bigleaf maple named Beatrice by my daughter, Claire, when she was a toddler, and continued through the woods, drawn toward the copse of American beech sited by Olmsted among grasses overlooking a wooded bluff and the Salish Sea, with glimpses of the Olympic mountains between branches. It is a favorite place to deploy my thermos of tea and read or write.

Departing the sheltered woodland and wandering across the meadow toward the beeches, I saw that one of the grove's eighteen trees had been cut down. I had been there the day before, so the loss was new. In the midst of the remaining beeches, I knelt next to the stump, which was surrounded by lengths of the trunk that had been sawed into pieces and scattered amongst the standing trees. I ran my hand over the crosscut and beheld the abrupt cessation of the spiraling growth rings—the stump reaching into the earth, confused and still-living roots visible among the grasses. The thicket seemed enfolded by a shroud of lamentation and bereavement, branches reaching and whispering. Then silent.

Humans, like other animals, feel and respond to prerational signals and senses all the time. We enter a restaurant, say, and no matter how pretty and candlelit and well reviewed on Yelp, as soon as we step inside we get an "I just don't want to eat here" feeling. We walk into a roomful of friends, and though all may be smiling, a troubled moodiness hangs in the air. Some days I follow a forest trail that I have walked hundreds of times and feel inexplicably overcome with discomfort; something is off— something I can't understand, explain, *or* ignore. I'll retrace my steps and find another path.

We flail in English with poor words for these moments. Words like *energy* or *vibe*. But in Japan I learned a word for this very thing—*fun'iki,* an ineffable feeling of goodness or badness or contentment or discomfort. *Atmosphere* may be the closest we can come.

Perhaps part of the Japanese language's ease with such an intangible concept has to do with another expression. Where I clumsily attempt in these pages to mingle notions of body, mind, and spirit, they are gathered in Japanese into one elegant word: *kokoro.* Language so deeply embodies our ways of thinking, it makes sense that through *kokoro*—not body or mind or spirit in isolation, but acknowledged as a whole—we can grasp the *fun'iki* of a place. There was only one way to name the *fun'iki* that enveloped the beeches surrounding their fallen sister tree that day. It was grief.

Truth and Trees

In the last century and a half or so, the dominant Western paradigm has become the only worldview across time and space to so overtly conflate scientific fact with a more expansive sense of truth. Perhaps our growing technologies have created a pretense of self-sufficiency, separating us from the transcultural myths that once wrapped us in a layered and living nature. Yet there are things we still apprehend beyond the requirement of data or even the possibility of data. Most of us have felt such knowing in our lives—an experiential mystery that lies beyond the quantifiable, replicable demands of good science, yet which we understand to be wholly true. We permit it in our poetry, our art, our literature, but remain tentative about allowing the numinous into our everyday understanding of truth. It is wonderful that the more science explores the intricacies of the wild earth, the more these two—inner knowing and outer fact—coincide.

It turns out that many of the things we have always intuited to be true of trees are also scientifically factual. Trees comfort, heal, speak, move, beckon, befriend. Trees listen, trees sing.* And though we have felt all of these things in the presence of trees—the solace, the communion, the rustling announcement of coming rain, the uniquely arboreal responsiveness—I doubt that even in the great sweep of such knowing we could have intimated the intricate, entangled processes through which all of these things occur.

This science was a long time coming, partly because of our preconceived divide among humans, other animals, and plants. Scientific-cultural entrapment has for so long lured us into viewing human ways of being as a measure for *all* ways of being.

---

* They do not, however, grow better when we play Mozart for them, no matter how poetic the notion, which was popularized in the 1970s. If trees and other plants *do* respond to music (which remains a possibility) and if they do prefer Mozart, they do not express it through increased growth or other measures we have yet recognized.

Responsiveness requires a nervous system. Sharing requires a mind that makes choices. Behavior requires a body that moves. Dancing requires feet. Affection and friendship require a heart—figuratively and physically. And any kind of intelligence at all requires a brain. In human-animal terms, trees have none of these things.

More than thirty years ago, when University of British Columbia professor Suzanne Simard was studying the understory of her Pacific Northwest ecosystem, she noticed something curious. Paper birch saplings were removed as "weeds" in Douglas fir plantations, assumed to be in competition for light and soil nutrients with the desired firs, which would eventually be logged for their beautiful wood. But instead of helping the fir saplings to thrive, this removal of birches resulted in the firs becoming less rigorous, growing more slowly, sometimes withering, or even dying. She witnessed a lab experiment in which pairs of pine saplings were grown in small root boxes. After one of the saplings was exposed to carbon 14, the radioactive gas was soon detected in the other sapling. The little trees were connected. *Sharing.* In light of these and other observations, Simard wondered: What if trees are not really in competition, as has always been the view in forest management? What if, instead, they somehow fortify one another? What if their flourishing is interwoven?

As a visionary young scientist, she was very much alone in the male-dominated discipline of forestry. Her proposed research was marginalized; grant money was denied. Yet she scraped together equipment, minor grants, and graduate students. She explored the underearth, the places we, in our terrestrial aboveground life, rarely pause to even think about. Now her body of research stands at the forefront of an entirely new understanding of trees, their responsiveness to one another and to the world, the forests they create, our capacity for relating to them, and our knowledge about how to protect them.

In the 1990s, research by Dr. Simard and others began to reveal that the vast underground mycelial tendrils of fungi (mushrooms are some fungi's aboveground fruiting bodies) already known to botanical science were something far more complex than anyone had ever imagined: the basis of a bustling constellation connecting myriad species. The research is extensive, decades long now, joined by many scientists, and still emerging.

While fungi were historically believed to have a parasitic, disease-causing, or decompositional relationship to flora, we now know that certain common fungi form a delicate affinity with trees and other plants. Cellular fungal hyphae gather to form a diaphanous threadlike tangle of mycelia, an eternal matrix woven throughout the soil—miles of mycelia in a square inch, and thousands of miles beneath each footstep. These adjoin and penetrate fine root systems to create a symbiotic relationship of mycelia/root/rhizome that has come to be called the mycorrhizal network, or "wood wide web."

Plants and fungi have been cavorting in this way for more than four hundred and fifty million years, forming an ancient mutualism. Mycelia invigorate roots with nutrients, including nitrogen and phosphorus, which they extract from the soil but which trees cannot obtain by themselves. The fungi, in turn, wrest the carbon-rich sugars that trees manufacture during photosynthesis and use it for their own growth. The circle turns once more as the mycelia assist root structures in extending throughout the undersoil, connecting trees to one another, allowing them to share a host of nutrients. And to top it off? The mycelia expand the reach of the electromagnetic pulses that are formed and transferred through the tips of trees' roots, forming forests of social connection, intercommunication, and the kithship of community.

Meanwhile, other research confirms that through chemical and olfactory signals in their leaves and branches aboveground, trees announce to other trees danger at the approach of leaf-feasting insects, news of coming storms, and times of safe repose. It is not only trees but likely all plant species—shrubs and ferns and grasses and mosses—that communicate in similar ways (though not exactly the same for each group or species). Dr. Simard recognizes all of this as a "kind of intelligence." Belying the myth that science and poetry are separate entities, she writes that trees forge "their duality into a oneness, making a forest."

Tropical forest ecologist and fungal network expert Merlin Sheldrake takes the breakdown of duality a step further in his wondrous book *Entangled Life,* pointing out the truth that discussing mycorrhizal activity with a focus on trees rather than fungi is plant-centric, in much the same way we historically ignored the responsiveness of plants and trees because of our human/animal centrism. It is perhaps an easier intellectual leap for us to think of trees as communal entities than to think of fungi as such; trees reside in our sphere—we can see them, lean upon them, nap beneath them. The fungi that invisibly tie all of this together are more elusive but every bit as essential and worthy of our attentiveness.

Dr. Simard's current Mother Tree Project will take a hundred years, following the individual lives and collective forests of Douglas firs, lodgepole pines, Ponderosa pines, and western larches at locations across Canada. (Mother trees, also called "hub trees" by Simard, are the most mature trees in a forest, hosting the majority of fungal connections. Within a society of trees, they offer the preponderance of protection through shade, shelter, nourishment, and communication.) Asked about the study's goals for an essay in *Smithsonian,* she said:

How do you conserve mother trees in logging, and use them to create resilient forests in an era of rapid climate change? Should we assist the migration of the forest by spreading seeds? Should we combine genotypes to make the seedlings less vulnerable to frost and predation in new regions? I've crossed a line, I suppose. This is a way of giving back what forests have given to me.

And what is that? "A spirit, a wholeness, a reason to be."

Growth Rings

The same constellation of communication tools that benefits trees' living also makes trees receptive to sudden severance, to loss. The beeches in my park are American beeches, *Fagus grandifolia.* Though their foliage is not particularly grand, their branches spread and commingle in a diffuse canopy, murmuring together in the wind. We know that beeches, spruce, oaks, and certainly myriad other species yet unstudied, register a version of pain—not human pain or dog pain, but tree pain. When an insect or caterpillar bites a leaf, electrical signals are released by the foliage tissue, mirroring the response of injured human tissue.

Neurotransmitters that we typically associate only with the demands of human life are found in trees and other plants: dopamine (in humans, it regulates emotional response and feelings of physical and emotional well-being), serotonin (mood and social behavior), and glutamate (communication between nerves, learning, memory). Trees tend daily to the details of living, biologically occupied with the very same activities as we are: growth, the gathering of light and water and nutrients, reproduction, care of young, maintenance of community, preparation for seasonal heat and cold, protection against disease, regeneration after injury, succumbing to inevitable death, feeding the soil into new life.

One day my friend Kersti, a tree expert who was experiencing an episode of depressive anxiety, said to me, "I wish I was a tree. They can never be alone." Like us—with us—they dwell in community, in kinship, in a kind of tree-friendship. A grove of beech trees ensures communal flourishing by sharing nutrients through the hidden mycorrhizae that help to equalize the strong and the weak. Trees feed one another. We know that certain species of trees support even sick individuals, sending them additional nutrients, just as we bring soup to ailing friends (and know that one day the giver may become the receiver). Some will rush to help even a stump.

Ecologist Brian Pickles, of the University of Reading in England, partnered with Simard and graduate student Amanda Asay to study Douglas fir seedlings in a recent project at UBC.

They found that the little trees recognize their relations, their root tips responding uniquely to their genetic kindred. How does it work? "We have no idea how they do it," says Simard. Trees stand outwardly as individuals but are joined into a radical reciprocity, so recently beyond the imagining of science. Trees reach for one another, invisibly, truly, beneath our feet.

Mary Oliver wrote a poem in which she lamented the inability of trees to move about freely, and it gave me pause. For one thing, in spite of being labeled "sessile" by botanists, trees and other plants move a great deal. In the 1990s, the BBC released a six-episode documentary, *The Private Life of Plants*, hosted by David Attenborough. Much of the discussion is out of date, but watching the time-lapse photography of botanical motion is a revelatory and almost psychedelic experience. Trees unfurl, stretch, grow, reach, follow the light, quieten in the dark. Most of this activity passes beneath our notice, as our own apprehension of movement-in-time is molded by the relatively speedy pace of our human motion in the world.

Might the trees, if writing their own poem, look at *us* with pity, instead of the other way around? So uprooted, meandering, separate, skimming the soil's surface, shielding ourselves from the night, our heavy limbs unresponsive to light wind. Lacking in fragrance, unvisited by birds, nothing nesting in our hair. Dying so young. Yet the trees, if we allow them, will enfold us in their grounded grace.

Communion and Kith

I have read that trees' interconnected frailty and attendant sharing of strength—the sort that cares for stumps—is possible only in undisturbed forests. But here in my urban woodland grove of just eighteen nonnative beeches, now minus one, I am uncertain. The trees stand in a field of grasses and tiny daisy-weeds. Some days trendy Eno hammocks or slacklines run between their thick trunks. Picnickers sip and nibble in their dappled shade. But most weekdays the park is sparsely populated and the trees stand serene in their idiosyncratic urban forest, where they have been maturing together for more than a hundred years. Their intertwining runs deep. How could they not lean with the rooted affection of tree-kinship over their friend, lying so suddenly upon the earth? In his luminous book *The Songs of Trees,* David George Haskell speaks to an ecological grieving that unfolds when a tree is lost:

> For the other creatures that depend on living trees, death ends the relationship that gave them life. The living tree's partners and foes must find a new live tree or they will themselves die. Much of the understanding of the forest that dwells embedded in these relationships also passes away. The trees' particular knowledge of the nature of light, water, wind, and living communities, gained through a lifetime of interaction in one location in the forest, dissolves.

And my own response? Like all trees (and animals, including humans), these beeches have a bioelectric presence. Our feeling for a place and its inhabitants, our sense of kithship and kinship, is strengthened and deepened with time and attention. Just as we know from a glance at the face of a beloved friend how they are feeling, we can become attuned to the *fun'iki* of a place we know well. I have written in this grove of trees, read poetry with my back against their trunks, pressed my face against their purple-silver bark. I have napped beneath them, my dreams of trees so detailed that I felt they were infiltrated by the beeches themselves. I have watched their seasonal quickening, flourishing, and release, season after season. I have visited these trees more days than not in the past twenty years. I was personally sad to be sure about the felling of this individual tree. But how far-fetched is it to wonder if my sorrow was not isolated but *shared* with the trees in the grove? Could I have been grieving *with* the trees, rather than simply among them? I do not know in fact, but I know what I believe to be true.

Forest Risk, Forest Attunement

The stump of the felled beech at my woodland park shows spalting—a sign of ill health that in traditional forestry would warrant removal. The disease that caused the spalting may, over time, have spread to other trees. Or it may not have. Rightness or wrongness of this particular tree-felling aside, city parks departments across the country are quick with the saw.

There have been a very few tragedies in wooded city parks—

people injured and even killed by falling branches — and so I do
not say this lightly. But this park and others like it are rare —
large enough to function as ecological systems with fragments of
old-growth canopy and an avifauna composed of resident and
migratory birds almost as diverse as that of woodlands far from
city borders. Such places offer sanctuary for all creatures, and
more than this — they are forest *entities*, with the full poetry of
essential connections seen, unseen, and even unknown. If we
want to maintain any semblance of urban forests, if we want to
offer a hint of rare and vital wildness, then risk (and the common
sense it demands) must be part of the whole. Wildness cannot
be fully policed. We are asked to enter the imperfect woodland
refugia that we have miraculously managed to sustain in urban
settings with a light step and with complex joy.

Weirdos Among the Trees

Dr. Simard suggests several things we human animals can do to
assist trees in their lives and forest-making, the most significant
being to simply spend time among them. Simard grew up in a
logging family and found her early inspiration as a child in British Columbia when she would lie "on the forest floor and stare
up at the crowns" of the ancient Douglas firs and western redcedars. It is in being with a *particular* forest that we come to understand its singularity, what it needs to flourish and thrive — and it
is how, in interbeing, we ourselves come to flourish and thrive in
response. Invoking one of the great spiritual metaphors, Thomas
Merton affirms this truth in his diary: "Living away from the

earth and the trees we fail them. We are absent from the wedding feast."

It is said that Coyote appears crazy in his dancing to those who cannot hear his mystical music. We have come to an earthen moment wherein we must make all the connections we are able with the whole of life, no matter how at-risk that puts our public-facing facade of normality. Look at the vapid homogeneity of the wealth-based, earth-denuding, dominant culture: is this the approval we seek? When we turn to the sweet, ragged edges of society, we see the people carrying violins, mandolins, pens, microscopes, walking sticks. The ones with ink on their hands, paint on their faces, mosses in their hair, shirts on sideways because they have been awake all night in the thrall of a new idea. This is where the art of earth-saving lies. We are creating a new story—one of vitality, conviviality, feralness (escape!), wildness, nonduality, interconnectedness, generosity, sensuality, creativity, knowledge of the earth and all that dwells therein.

On walking barefoot, Thomas Merton wrote, "Perhaps this might appear quixotic to those who have forgotten such very elementary satisfactions." The superficial pastime of caring what others think of our love for the earth is long past, and subservient now to the "elementary satisfactions" that stitch us to a troubled planet. With this in mind, I offer some other-than-normal but beautiful practices for bringing reciprocity to the trees among us and the forests they create.

Return, and be claimed. Choose a tree to sit with. This practice intertwines with the "still spot" of animal relating. Visit this tree every day for a season, a year, a life. From this simple practice, everything is learned. While rooted in place, trees gather all things required for their existence—and much of our own. I know that while I cannot claim land, or beings upon land, there is a sense in which I can feel claimed by them. Lynda Mapes studied a single oak over the course of a year in Harvard Forest, the university's 3,000-acre research area. She watched, sketched, researched alongside scientists. She climbed high to the oak's crown, she napped among the oak's branches. Lynda wrote about the experience in her book *Witness Tree,* and told me she has no doubt that in time, the tree recognized her presence— that they belonged to each other. To return and return and return is to come into relatedness with a specific tree and the surrounding land. It is the great lesson.

Bodies touching. For beloved Buddhist monk Thich Nhat Hanh, hugging is a spiritual practice, one that extends to all beings—people, animals, and trees. "When we hug, our hearts connect and we know we are not separate beings." Nhat Hanh contemplates those whom we hug:

> You have to make him or her very real in your arms, not just for the sake of appearances, patting him on the back to pretend you are there, but breathing consciously and hugging with all your body, spirit, and heart... "Breathing in, I know my dear one is in my arms, alive. Breathing out, she is so precious to me."

If you do this, the tree being hugged "will be nourished and bloom like a flower." Is someone around to see you hugging the tree? This is the part about not caring what others think. Tree-hugging dirt worshipper? Thank you, yes.

Entrainment. With shoes shed, we become open to the beneath-ness cultivated by trees, and we draw closer to them as our mentors. Putting not just a cheek but an ear directly upon a tree's bark allows deepest listening. The rhythmic movement of the blood in our bodies, the craniosacral fluid of our spines and skulls, mirrors the movement of sap in the trees. The breeze in our rustling hair is the same as that in the leaves, the branches, the clouds. The longer we stay, the more entrained—indistinguishable—these rhythms become.

Reading to trees. Trust me on this: thinking about reading to a tree feels stranger than actually reading to a tree, especially once you get going. The other day, I walked to the woods with a particular cedar in mind and a book in my pocket. What does one read to a cedar? Though I am sure there are many possibilities, the answer on this day for this tree was obvious: Emily Dickinson.

If you read aloud to a tree, you will find yourself in the company of an uncommonly attentive, responsive (possibly even grateful) listener. And is the tree's reaction—a singular stillness, a branching rustle, the dropping of golden leaves after a particular line,

The Brain—is wider than the Sky—

as if to whisper, *Yes, yes!*—real or imagined? I am not at all certain that it matters whether we form a syllable in answer to this question. We are always, and everywhere, part of a graced conversation.

It is almost impossible to pass by a tree with whom you have shared such intimacy and not feel beckoned as you pass by in the future. In his stunning novel *The Overstory*, Richard Powers draws on ancestral knowledge from cultures across the earth when his character Adam wonders aloud whether his friend Olivia, who died acting in defense of ancient forests, could really hear trees speak, as she had claimed. "Trees used to talk to people all the time," he answers himself. "Sane people used to hear them." The silent narration continues: "The only question is whether they'll talk again, before the end."

Sane people still do hear trees speak; it is a matter of making ourselves available to listen and, when we are ready, to respond. These essential conversations invite respect, openheartedness, open-mindedness, a decrease in our tendency—even our ability—to commodify another's existence. They broaden our sense of living within an inspirited, sacred world. This is the rooted foundation of all bold, outer activism on behalf of wild lands and life.

GO TO THE TREES

Create

The Art of Earth Activism

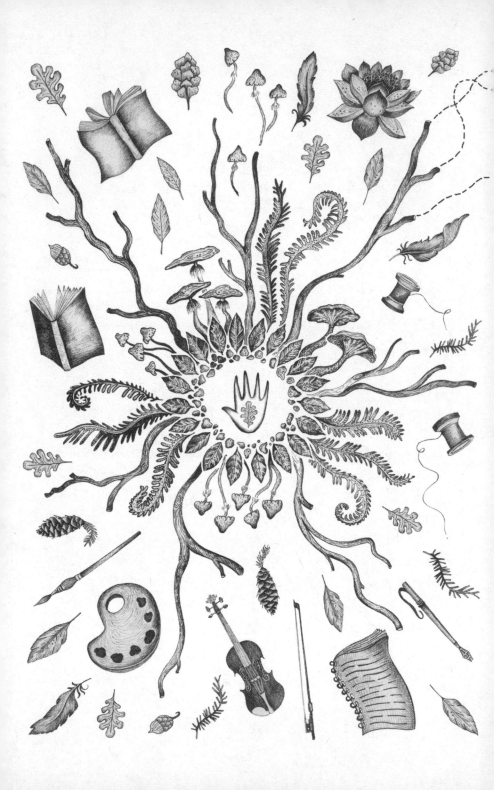

"The fox knows many things, but the hedgehog knows one important thing," proclaims an ancient fragment attributed to the Greek poet Archilochus. It is implied that the hedgehog curls into the side of higher wisdom, and we are advised to sink into the still depths of this hedgehog way of knowing. Discerning people have told me that because no one can do everything needed to help the troubled world, it is most impactful (and least confounding) to focus our passion on one thing. Pick one cause, one creature, one ecosystem. Go deep. I think this is mostly true.

But in my favorite story from Saint Thérèse's diary, the book from my childhood Easter basket, one of the older girls in the family passes on to her sisters all the bits and bobs she uses to make doll clothes—scraps of ribbons, threads, and cloth. She places these things in a basket and offers them first, as a courtesy, to Thérèse—the youngest—who, not courteous in the least, plunges her chubby little hand into the basket, grabs everything, and says, "*Je choisis tout.*" I choose all.

Whenever I pick through my own basket of tangled embroidery thread of every color on earth and the tiny golden crane scissors from Germany that my mother gave me when I was little, I think of this story. With these simple materials I have adorned and mended clothes, of course, but also: my tent during a storm; the wing of a migratory warbler; a friend's broken heart. More.

The One Thing and the Ten Thousand Things

It is hard now. In this anxious time on our planet, we are well past the *50 Simple Things You Can Do to Save the Earth* mentality of the early '90s, when the general public knew just a whisper of what climate change and the rampage of continued wild habitat destruction would mean. We know now that there are ten thousand things and more, some of them simple, some of them things we have never done before, some of them barely possible, every one of them indispensable, all of them requiring our attention, our service, our love, our action. *Reduce, reuse, save, eliminate, meditate, give, work, howl. Orca, cedar, redwood, wolf, caribou, frog. Coral reef, tropical rainforest, temperate rainforest, desert. Air, earth, water. Cultures, languages, tribes.* An endless, sacred litany. Nothing can be removed from the list. Every countless thing essential.

No wonder we feel such paralysis. And yet we are called to act anyway, within this paradox of the many and the one. In this time, perhaps we do not choose between the fox and the hedgehog but embrace them both. No one can *do* all things. Yet we can *hold* all things as we trim and change our lives and choose our particular forms of rooted, creative action—those that call uniquely to us.

In Western mysticism there is the beautiful idea of the "communion of saints." These are not the saints ordained by an institutional religious process; they are all of us. In the communion of interconnection we do not individually possess all gifts, but partake of one another's gifts as we offer our own, forging a graced and complete whole.

When I lived in Kyoto, one of my favorite destinations for day-dreaming was Sanjūsangen-dō, the temple of the Thousand-Armed Kannon—a version of the feminine bodhisattva of compassion I spoke of in this book's invocation. Here, there are a thousand golden human-size statues of Kannon, each with forty-two arms, forty-two hands. Each hand offers a different gift to the thirsting world—a lotus blossom, a sword, an eye, a bird.*

In the long dark hall, the wood floor has been polished smooth by the bare footsteps of millions of pilgrims over nine hundred years; it is always cool, even in the sweltering Kyoto summer. The endless figures are beguiling and uplifting. I would wander among the forms when I visited, feel the thousands of fingertips trail through my hair, patter upon my body, hear the voices whispering—sacred things, difficult things, the words indistinct

---

* The thousand arms per statue proved too much for the confines of this temple, though it has been accomplished on single statues elsewhere. When you discount the usual two arms, the remaining forty multiplied by the twenty-five planes of existence of the Tendai Buddhist sect bring a metaphorical thousand arms to each figure.

but luminous. I would leave with arms sprouting from my own back, hands open to receive, palms full of blossoms to share— the petals flowing behind me, I knew, as I walked. Hedgehog, fox, one heart, one thousand arms.

Create-Creature-Creation

The words arise in a tangle. *Create* is from the Latin *creare,* to cause to grow, *to bring forth,* and passed via Old French to the English, *creature,* a *being* brought forth. All beings, all creatures, are coformed. Like the earth and all that dwells herein, we are made of clay, of water, of blood, of stardust.

Carl Sagan is famous for crooning, over and over, that we are star stuff. He has been mocked for his ponderous intonation, but he was entirely correct. Dr. Ashley King, planetary scientist and stardust expert (an enviable job description), states: "It is totally 100 percent true: nearly all the elements in the human body were made in a star and many have come through several supernovas." *Oxygen + carbon + hydrogen + nitrogen + calcium + phosphorous + potassium + sulfur + sodium + chlorine + magnesium = star-human.* The stuff of the cosmos is woven into our bone branches and wanders in our blood rivers.

"The individual is the meeting place of the four elements," writes John O'Donohue of the ancient Celtic perspective. For humans, this is a hallowed opportunity: "We have come up out of the depths of the earth. Consider the millions of continents of clay that will never have the opportunity to leave this underworld. This clay will never find a form to ascend and express itself." Yet we have this privilege, and with it a sacred obligation to live with meaning. The word *obligation* sounds burdensome but is rooted beautifully in the Early French *ligament*—that which binds us together, like a braid of blessing herbs, like a bundle of logs for a warming communal cookfire.

Years ago I would have avoided the word *creation*, which has been dulled by the unreality of fundamentalist creationism. And yet all of the other words we so often deploy to avoid the murky waters of creationism—*world, natural world, earth, biotic life,* or even just *life*—fall flat next to the poetry of the word *creation*, implying as it does an unending ground of existence, a matrix of constant renewal. *Creation* embraces the so-called nonliving substrate of life: things watery, geological, cosmological, mineral, and atmospheric ("so-called" because there is no sensible way to disentangle the living from the nonliving).

Even time breaks down within this calculus of interbeing. We stand in a spiral— rather than a strictly linear—continuity with our ancestors and the ancient cosmos. We still see the light of the stars that died long ago and that now form our living bodies; so, too, do our actions reach into the future of all life and death. It matters what we bring forth with the matter of our bodies. We create, as cosmos-formed creatures, within creation.

The Creating Wind

Our elemental bodies are the vector of action in the world, but the inspiration that moves us to act has long been associated with the most invisible of the elements. Life/spirit/breath twine inseparably in the roots of this word. *Inspiratus* is the creative wind, the divine breath. And to *be* inspired? To be inspired is to be breathed upon. Walt Whitman sang,

> My *respiration and inspiration, the beating of my heart,*
>   *the passing of blood and air through my lungs,*
> *The sniff of green leaves and dry leaves . . .*

To listen for our place on earth is to be inspir*able*. Inspirability is essential, coexistent with the creative act.

Clarissa Pinkola Estés describes the presence of *el duende* in certain Latin folkloric tales as a mystical substance, the wind that breathes its *inspiratus* upon us. It is in rooted contact with the natural world that we come to hear this wind, to let it touch us, to breathe it in, to act from its voice. "When the leaves of the trees shake," Estés writes, the people say, "Ah, *el duende*." *El duende* is the force felt by all mystics, by all artists, by all of us as creative activists when we let our actions be fierce, healing, blessing, protecting, loose, and windblown in the world.

Living is itself a creative act, and the art of living matters in all spheres. The demarcation between enmeshed concepts of creativity (the making of a sculpture, the writing of a book, the painting of a still life, the composing of a quartet, the penning of a poem) and activism (the joining of a protest, the marching for a cause, the laying of one's body before a bulldozer, the willingness to be arrested for the right to clean water) is a false one. All of these can embody both creativity and activism. Zadie Smith spoke in an NPR interview about her sense of conflict over being a writer within a friend-community of traditional activists who are ever marching, megaphoning, organizing. She sometimes feels passive, but recognizes over and over again that her writer's question "Yes, but why?" presses the seemingly more active work into the broader discussion and toward structural change. The ways and tenets of rootedness prepare us to share our unique passions—whatever they are—on behalf of a beloved, suffering world. Our work is to allow this passion to affect our existence, to let the inner ecology of our lives come to touch the outer ecology of the earth. *This* is the creative art of earth activism. The human task now is to bring it.

The Science of Creatures Creating

It is not easy to pinpoint the ways that nature underlies our creativity. The human response to the natural world is entangled with our conscious and unconscious minds, our imaginations, our spirits. These are difficult things to access for traditional scientific analysis. How does one quantify a *feeling* of expansiveness or creativity? And yet for decades there have been attempts in the fields of environmental psychology, ecology, and psychiatry to furnish empirical evidence for this dimension of the human relationship with nature. Papers from the '70s and '80s by Theodore Roszak and others suggest that nature is important not just for material needs but also "for psychological, emotional, and spiritual needs." Physician-ecologist Bryan Furnass proposed that the experience of nature bolsters neurological activity of the brain's right hemisphere, restoring "harmony to the functions of the brain as a whole." While researchers continued to expand on this theory over the following decades, and though there is a mountain of analysis detailing how nature might enliven our creative minds, it wasn't until quite recently that the specifics of how this might work began to coalesce.

In a 2012 paper, Ruth Ann Atchley and Paul Atchley, of the University of South Florida, and David Strayer, of the University of Utah, presented their summarized years of thought and research on the psychological and cognitive effects of time spent outdoors. In one study, hikers were divided into eight groups of about seven hikers each and sent off, sans cell phones, into the wilderness for a relatively strenuous four-day backcountry trip. None of the participants knew the purpose of the study, or that other cohorts were involved. Half of the groups were tested before they set out, using the respected Remote Access Test (RAT) for assessing higher-order cognitive function; the other half were tested on their return.

The RAT measures our ability to identify connections between strings of related words (blue, cake, and cottage all connect to cheese). This task calls upon higher-order cognitive ability, insight, and capacity for imagination. Since the '70s, when these were vaguely known to be right-brain activities, we have come to understand that such creative functions lie in the prefrontal cortex of the brain, the same part that is overtaxed by the constant demands of our modern, tech-heavy environment (the human brain, it turns out, reacts similarly to the ding of a text message or the wail of a siren as it would to the growl of a bear behind us in the woods: as a *threat*, demanding a constant state of alertness). The returning hikers scored 50 percent higher on the RAT, showing elevated creative capacity for the kinds of higher-order tasks in the study.

A limitation of the study, coauthor Strayer readily admits, is that it does not tease out the positive effects of the natural environment from other aspects of the experience that might lead to a rejuvenated prefrontal cortex—distance from cell phones, for example, or the joy of physical exertion, or just being away from the rigors of daily life, no matter where "away" is. But Strayer feels that this paper, alongside other international work, bolsters the idea that the relaxed attention we cultivate outdoors is key.

There is plenty of new research to bolster Strayer's work. In a recent study at Heriot-Watt University, Edinburgh, Professor Emeritus of Environmental Studies Peter Aspinall and his colleagues employed mobile EEGs to monitor the brains of participants as they walked between built urban spaces and green urban parks. Across the board, the EEGs indicated higher levels of frustration, distraction, and general arousal when in the built environments, and a more meditative state when in the green spaces, providing support for what Strayer calls "the brain default network" that is linked to creative thinking.

Such outcomes are not a surprise. Many of us turn intuitively to the outdoors for renewal of mind and mood. The wonder of this research lies in the particular details of how the creative brain functions. Yet no matter how much we know of such things, there will hover something more—the *inspiratus* that both participates in our material being and invites us to sing beyond it.

The Creative Ecology of Home Economics

I will venture that most of us reading these words live within a modern economic story of so-called privilege; our needs are well met, and constantly so. We know that this is not true in much of our own country or in most of the world, where human and beyond-human life struggles to meet the basic necessities of sustenance, shelter, and health. We are unnaturally comfortable.

I write from a house heated to a sensible 66 degrees at all times, and with a refrigerator stocked full of "necessities" like chardonnay and Greek olive tapenade; a shard of artisanal dark chocolate sits ever-ready next to my shiny MacBook Air for moments of supposed need. When I go camping, I take more with me than billions of people on earth own in total. And no matter how many times I hang the laundry on a clothesline instead of popping it into the dryer, I will never make up for my airplane travel. I speak from entitlement that I have done nothing to deserve, knowing that any step I take toward economic and material simplicity is cradled by a great safety net within my social class and my family, to which I can return at whim.

Progressive as popes go, Francis wrote in his encyclical *Laudato Si': On Care for Our Common Home* that our "green rhetoric" is framed by professionals, thought leaders, organizations, and media who reason from a place of comfort within "a high level of development and a quality of life well beyond the reach of the majority of the world's population," which can lead to "a numbing of conscience and to tendentious analyses which neglect parts of reality."

Still, we are wired for some measure of uncertainty in our comfort level, and when we do not have it, we seek it: we carry what we can uphill on our back to deploy small camps; we push ourselves to uncomfortable exertion on trails, or in athletics; we fast for spiritual equilibrium. It is time to bring a measure of this discomfort into our daily lives in order to stand with all of life. But how? With what?

Part of the answer is close to hand; our everyday lives at home offer the dearest and most potent form of creative activism. There remain places in the remote wilderness where no human has ever set foot, yet there is no place on earth untouched by human presence, as climate crisis teaches. Lynda Mapes notes in her award-winning series for the the *Seattle Times* on the southern resident killer whales (the community of the mother Tahlequah), who live primarily in Puget Sound, that the main thing keeping this threatened population from the robust health exhibited by the nearby northern orca communities is that they share their home waters with me—and the six million other people of the greater Seattle region. They thereby share the residue of our affluence that flows into their waters, affecting every cell of their bodies and contributing to the devastation of the salmon who are the southern resident orca's sustenance.

The activities of our households—the ways we eat, clothe, and warm ourselves, the ways we move about, the ways we dispose of our shocking accumulation of waste—bind us biologically and spiritually in ecosystemic relationships far beyond our doorstep, into the farthest reaches of the wildest places. Separateness is dismantled in our home economics. Our lives are coyote lives, cougar lives, oak lives, orca lives.

Rebecca Solnit writes in her essay "Our Storied Future" that "as citizens engaged in the daily task of remaking the world, we get to choose our stories—the stories that divide and conquer or those that tie things together with possibility." This is not about reinforcing what Edgar Villanueva refers to in *Decolonizing Wealth* as a "colonial division" in modern philanthropy, where there remains an imposed gulf between "Us vs. Them, Haves vs. Have Nots." It is about pressing ourselves more strenuously than ever to bring our actions in line with our hearts, to create a new story of just and sacred economics, and when able to take the radically countercultural step of *deleveraging* our bank accounts and our acquisitions. It is about rethinking the food we choose, and the attitude with which we eat it. About sharing what we have within the spirit of our relatedness. I do this in small ways; I want to do it in profligate ways. Perfection is impossible—I will fail a million times in this endeavor every single day. Yet together we go forward. Moving toward a radical simplicity in these things is an act of bringing forth a more intentionally congruent life, a creative life beyond overt cultural narratives and expectations of society and family, in solidarity with all humans, all beings, and in reciprocity with the dreaming earth.

Name Your Charism

Charism is an enchanted word from the Western monastic tradition. A charism is a particular gift or power—almost a superpower, received in grace through *inspiratus,* the spirit-wind, and returned to the world as "favor freely given."

Each monastic order has its own charism. Benedictines live under the watchwords *ora et labora*—pray and work—and share the benefits of these practices in their charism of radical hospitality, the belief that all guests are to be welcomed as the Divine. Cloistered Carmelite orders, like the one Thérèse joined in nineteenth-century Lisieux, were contemplative, offering the gift of silent and sung prayer as a mystical devotion within the world, weaving a web of sacred interconnection with life beyond convent walls.

Thérèse herself admitted to sleeping through most of these communal prayers but evolved her own personal charism: doing every small thing in her life with fullness and love. "To *pick up a pin* for love *can* convert a soul," she wrote from her innate mystical wisdom. She knew that within her provincial French convent she could not be the activist and reformer that her namesake, Teresa of Ávila, had been in the sixteenth century. But before she died of tuberculosis at the young age of twenty-four, Thérèse created—and lived—a philosophy of spiritual freedom, very much like the Zen tradition (of which she knew nothing), bringing wholeheartedness to every action for the benefit of all, like a bodhisattva, while remaining unattached to the outcome of her "little works."* An activist dimension arises within this charism of everyday mysticism; to bring generosity and simplicity to

---

* The Carmel of Lisieux has been much romanticized since Thérèse was sainted and became a Doctor of the Church. But the monastery's deplorable conditions surely contributed to her early death. Food and firewood went first to the priests, then to male monks, while the sisters shivered in dampness and hunger beyond the austerity of the monastic rule that guided the men's daily lives.

every encounter thwarts an economic-political paradigm based on rampant consumption. The heart and life of individual agency has everything to do with broader systemic and ecological change.

We each have our own calling, our own charism—the unique gift of our own life and creativity. It is useful to think of it this way: as the particular grace that is ours to give freely, with devotion, with service, with joy, though not without hardship. As Clarissa Pinkola Estés warns about the creative spirit of *el duende,* "If you think it costs nothing to have it, all your hair will be burned off."

Rachel Carson experienced this as an author. She heard a personal calling to be a voice for the sea in her first three books, then for birds and all of nature in *Silent Spring.* And she knew that for her the way to express this charism lay in her singular creative gift—her pen, her writing. Our individual charism comprises these twin dimensions: the earthen need that summons us, and the gift through which we bring our response to this summons into the world.

Rachel wrote to her beloved friend Dorothy as she was beginning her third book, *The Edge of the Sea,* that she was afraid to begin again, aware always of what she called "the crushing burden of creative effort." And yet she embraced her charism and returned it wildly. She told Dorothy that "the heart of it is something very complex, that has to do with destiny, and with an almost inexpressible feeling that I am merely the instrument through which something has happened—that I've had little to do with it myself."

She refers to "destiny," a pretty word that usually has a positive connotation yet etymologically relates not just to a sense of being called, but also of being *doomed*. Ha! Who has not felt it? Rachel is offering the gift that, if she had refused it or abandoned it or taken it up halfheartedly, would have made her life easier. But the idea of turning her back on the beautiful doom of her gift-obligation would make her quake with restlessness; it would wake her with insistent whispers; it would haunt her like a ghost.

Nothing has prepared us for this planetary moment. There is no prewalked path to offer direction. Our individual charism cannot be prescribed, proscribed, or even thought up in our head. It can, however, be listened for — a rooted, ongoing, reciprocal conversation with the wild earth — a spiral of inward, receptive stillness, and outward, creative action.

It may be painting, singing, writing. It may be tree sitting, anti-fracking demonstrations, fundraising. It may be tending: the garden, the farm, the wisdom of the elders, the faith of the young. It may be prayer for a threatened grassland, knowing the ways of the birds, beholding a coyote's gaze, sacred silence at the base of an ancient tree. It may be weeping for the loss of a forest, tithing a portion of our life-energy in the form of money for orca protection. It is most likely a combination of such things. It is certain to be countercultural and counterproductive according to all usual measures — those we have been taught, those that can be labeled with a sum or a number.

Walt Whitman asked that which we all primally ask: *The question, O me! so sad, recurring — What good amid these, O me, O life?* We hold the ten thousand things of the troubled earth in mind and in spirit while offering the few beautiful things that we, and we alone, are able to offer with our ten little fingers. Whitman answered his own question: *That you are here—that life exists and identity, / That the powerful play goes on, and you may contribute a verse.* This is all we have—our life and what we give. In our ragged wandering with padded feet and pricked ears and rewilded minds we find gifts from the wild earth, and we come to share our own gifts in return.

CONTRIBUTE YOUR VERSE

Spiral

A Wild Return

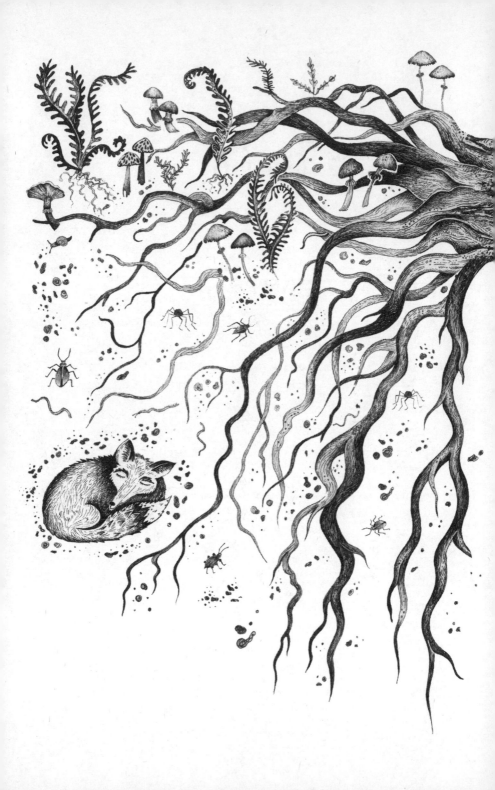

Time in nature is time in the presence of essential death. Leaves to soil, lemmings to owl food, deer corpse to winged vulture flesh, and our own bodies one day to...to what? This is my contemplation as I write, seated on earth, notebook in lap, back leaning against the trunk of a western redcedar. The tree lived for more than two hundred years before falling here, a few minutes' walk from my home. She has lain here for a decade, maybe longer. Her roots are tangled and exposed.

Yet she is not dead. Western redcedars become the most fertile of mother logs, their layered, ochre wood-bodies easy purchase for mosses, whose rhizoids sift the thready bark into a kind of tree-soil where ferns grow, and huckleberries, and tiny faerie versions of western hemlocks. Her presence here as simultaneous tree, trunk, soil, and seed delimits the duality of past and future. Over the years, the tree will sink further until she becomes earth, and the young she has suckled upon her body will continue to grow higher and more fruitful, spreading their own seeds and, eventually, their own bodies, upon this forest ground. Life, death, and life again.

The more time I spend lying upon the forested earth of my home ecosystem, the more I am conscious of what I have come to call *beneathness*—the writhing substrate that cradles every step we take, alive with decomposition below soil as the bodies of trees, leaves, and every creature are reclaimed for the nourishment of life above soil by decomposers, worms, fungi, beetles, termites, detritivores of all sort. Desert, oak, ocean, prairie, and steppe ecosystems all turn within their singular spirals.

Skin and Soil

New science supports a cellular intelligence and intricate ecology existing long after the seeming death of an organism. At the Southeast Texas Applied Forensic Science Facility, in Huntsville, researchers who study the decomposition of human bodies find that, far from being "dead," our bodies, when they are allowed contact with the earth, lie at the center of a convivial "cadaveric ecosystem." Most of us prefer not to think too vividly about what becomes of our bodies after we die, but it is an ancient and noble contemplation.

For hundreds of years, Buddhist monks in Vietnam have meditated in open cemeteries, bodies of their brethren in different states of decay around them. There, they envision the same processes that will inevitably be at work upon their own bodies. Saint Benedict, a fifth-century Italian monk and scholar, authored a famous book-length *Rule* for living that is still followed by many monastic orders and their oblates (including Benedictines, the Trappist order that drew Thomas Merton to monastic life, and even some Buddhist monasteries). Sisters and brothers following the *Rule* are counseled to "keep death ever before you." Such meditations call us to acknowledge more fully the insistent ephemerality of our existence and to live with more intention, generosity, humility, and love.

The researchers at the Huntsville facility, like the Vietnamese monastics, carefully study bodies in all phases of decomposition. Soon after our hearts stop, our bodily ecosystem consists mainly of the bacteria that live upon and within us—it is a vast number, with every surface and nook inside and out offering habitat for specialized communities of microbes. Thousands of species, trillions of bacteria. Soon after our cells become deprived of the oxygen provided by our beating hearts, the acidity of our cells increases and a chain of chemical reactions is set in motion. We begin to "self-digest," as cell membranes are consumed by enzymes, and their contents begin to spill out in microscopic rivers.

In 2014, the first paper to explain this process in depth was published by Dr. Gulnaz Javan, of Alabama State University, in which she named this small ecosystem the *thanatomicrobiome,* from the Greek *thanatos,* for death. While alive, our immune systems keep microbes out of our internal organs, but this immunity wanes upon death, and microbes quickly overtake our guts and our intestines, and then surrounding tissues, digesting them from the inside outward, using the leakage from damaged cells for food as they spread to capillaries, lymph nodes, liver, spleen, heart, and brain.

This colonization by microbes, along with the self-digestion of cells, allows bacteria to escape from the gastrointestinal tract, leading to putrefaction and the breaking down of soft tissues into liquids, gases, and salts. Now there is a move from aerobic bacteria, which require oxygen to survive, to anaerobic bacteria, which do not. (Bacterial organisms pass gas, and these "bacteria farts" are in part responsible for a body's developing stench.) Eventually, all of these gases and liquefied tissues are purged. Sheets of skin slip away, the body bloats. And while this certainly looks bad for the body, it is at this point that things get really interesting in terms of our physically nourishing life beyond our own thin husks.

When our bodies are liberated from the contours that shaped us in life, we become an ecological hub for a variety of insect species, many of which have evolved to live their entire life cycles, from egg to flight, around a decomposing human body. These insects in turn provide sustenance to a variety of other arthropods, reptiles, and birds. Different species feed at sequential stages of decomposition, giving rise to the modern science of forensic etymology—determining the time of death by the insects that are living upon a body. These insects also signal our presence as nourishment to vultures, ravens, and meat-scavenging mammals. Thus we might be carried by feather or paw to sky, forest, or mountain ledge.

Other recent science shows that cells do not die immediately but take hours, sometimes more than a day, to quiet and begin the transformation to this cadaveric ecosystem. Inexplicably, some genetic material contained within our cells does not even express itself until *after* what is medically termed to be bodily death. By the humanistic selfish-gene version of evolution, this doesn't make sense, and scientists have come up with no reason that it would be so. I have read in scientific journals that such genetic expression has "no purpose," though more likely it serves a purpose. It likely serves a purpose we do not yet understand. All of this is of value in tracing the tragic loss of loved ones, but it carries philosophical and spiritual meaning as well.

While working on this book for several weeks in the Santa Ynez Valley in California, I was bewitched by the oak ecosystem, so different from the green, mist-laden forests of my home. One warm day, I was on a long ramble with Kate McCurdy, the reserve director. After several seasons of drought driven by climate crisis, many of the valley oaks standing here and there in the windblown meadow grasses were brown with near-death — far more than there ought to be. "They say it takes a hundred years for an oak to grow," Kate pondered aloud, "and a hundred years for an oak to die." We walked in thoughtful silence, contemplating what this means for oaks, for us, for life.

On a recent family camping trip to the Olympic mountains, I took some time to visit with an ancient cedar mother log who is returning to soil, grown over with young hemlocks, huckleberries, and ferns, and blanketed with sphagnum moss — much like the mother log near my home but everything thicker and deeper and more extravagant in the botanical abandon of the temperate rainforest. I shed my shoes, climbed upon the tree, lay on my back, and closed my eyes, letting my body become heavy. I wondered how long it would take to feel a sense of mossy rhizomes reaching into my skin, the first loosening of bark-into-soil. It turns out not long at all — unsettlingly not-long.

Dying-as-spiral, with no set moment of absolute death, is in some way true for all organisms — cellular death is a long process, with unknown holistic value to body and spirit. My hope is that my own body will be allowed to take its time in dying fully one day, with all that might entail.

The List

This sensibility informs the to-do list I shared with my long-suffering husband:

1. Brain dead? Don't unplug me. At least not right away. Give my body time to settle into her loss of brain function. Go home. Read *Anna Karenina*. Watch all seven seasons of *Buffy the Vampire Slayer*. Vacation in Paris. Don't forget to pray for a miracle. After all of this, revisit the idea, and do what you think is best.
2. I am leaving you with the number of a death midwife I trust. If you are up to taking care of my body yourself, she will help you. If not, that's OK. But call her anyway and learn how to make sure I am not embalmed. No embalming.
3. If we still live in Washington, bury me at the natural cemetery in Ferndale, preferably close to tall trees.
4. For burial, I would like to be wearing a hand-sewn organic cotton or linen dress. Something simple yet flowy.

Wait, what? Poor Tom. He was doing fine, mostly, till we got to number four. And I realized, *Of course, how will my grieving husband figure out the creation of a cotton dress? Let alone organic and hand-sewn?* "Don't worry," I told him. "I'll make it myself." And I did.

Dress for the Occasion

Before I began the dress, I consulted the directors of two natural burial grounds in Washington State. In natural burial, the body is wrapped in a shroud, perhaps laid in a simple casket of biodegradable material (willow, untreated wood, woven grasses), and buried directly in the soil. There is no toxic embalming fluid, no thickly lacquered coffin lined with synthetic fabric, no concrete vault for interment—none of the practices that keep our dear bodies unnaturally separate from the cradling earth. The Order of the Good Death, an organization that works to emancipate the process of human death from both cultural silence and commercialism, makes it surprisingly straightforward: "For most of human history, what we now call natural burial was simply called 'burial.' A simple, shallow hole dug into the earth, and the shrouded dead body placed into the hole."

We all do the best we can in the cultures that we have been born into, and discussions of alternatives to current norms are in no way a judgment upon the choices any of us have made or will make on behalf of ourselves or our loved ones. Let us be at peace with all such decisions. It is beautiful that our relationship to death and care of bodies after death is evolving, but there is no one way to participate in this re-storying. In her inspired and challenging memoir, *Bless the Birds*, nature writer Susan Tweit chronicled her husband Richard's decline and death due to a brain tumor. Ever an educator in life, it was Richard's wish that his body become a teaching cadaver for medical students. After his time teaching post-death, his ashes were returned to Susan. When I asked her about it, Susan said, "I'd always imagined our bodies 'going to ground' together and gently moldering back into earth, but if it gave Richard peace to donate his body, that was his decision." As Rumi sang, "There are a thousand ways to kneel and kiss the ground."

It is increasingly common for traditional cemeteries to set aside corners for green burial. Other entirely natural cemeteries are in remote settings—often located within woodlands, meadows, mountainsides, and parts of intact ecosystems. In such burial grounds, carved gravestones are absent; the place of burial is marked by an ordinary stone, or nothing at all (GPS locations are kept on record).

When I spoke to the keepers of natural cemeteries in my state, I was told that it would be fine to be laid to rest in a homemade dress, though my loved ones might also like to wrap me in a shroud, or place me in a willow casket, or both. All materials, including the thread used to sew the garment, must be free of synthetics that would impede the process of decomposition. I chose a simple cotton muslin and used a pattern by designer Tina Givens. I was always afraid to try making one of her dresses because on her website even the willowy models look a little bit like potatoes in her flowing designs. But surely a dress settled around my still body will not evoke potato-ness? And even if it does, I doubt I will mind.

Hand-sewing is calming to me, and I chose to stitch my dress entirely machine-free. For whimsy and inspiration, I've selected some thread in a pretty shade of moss-green and continue to embroider quotes along the hems as fancy strikes. Joy Harjo: *Remember the earth whose skin you are.* Walt Whitman: *I bequeath myself to the dirt to grow from the grass I love... Your very flesh shall be a great poem...* Now and then I wear the dress on a forest walk, letting it become accustomed to roots and soil.

If any of these practices and ponderings sound glib or overly lighthearted, know that they are defense mechanisms. Naps upon decaying trees. Sewing of shrouds. Skulls of birds and coyotes enshrined as *memento mori* on the shelves of my study—I contemplate them daily in the palms of my hands, their intricate post-purpose: *Remember.* All of this is an attempt at a reckoning with the end of my own life, the constant presence of an inevitability I am as yet unable to fully brook. Some say peace with death descends upon us as we age, and perhaps this is so. For now, I struggle and I stitch.

An Earthen Immortality

The correspondence between Rachel Carson and her most beloved friend, Dorothy Freeman, spills over with their shared sense of wonder, mystery, and solace found in the things of earth, sea, and air—the dancing language of an inspirited world. In periods of extreme trial, they would from time to time attempt to comfort each other with traditional religious words. "May the Lord bless you and keep you, while we are absent from one another," Rachel wrote in a particular letter, knowing Dorothy was anxious about Rachel's upcoming cancer treatment. There is an occasional keeping in "thoughts and prayers." But such expressions had an awkward, out of place ring within the correspondence as a whole, and the friends made no effort to sustain them, even as they wrote extensively of "things of the inner being."

Yet as Rachel's cancer progressed, there was a growing urgency to make sense of what they believed. When the inevitability of Rachel's impending death became undeniable, Dorothy, who had always been agnostic about an afterlife, found herself in the dearly human position of embracing a belief in immortality, and wrote to Rachel with excitement about her vision:

Darling, I must tell you something that happened to me last night. (You said, you know, that we should not hesitate to share all our thoughts.) Suddenly, as I was lying awake in the darkness, a thought, that was so different, hit me like a revelation. To state it very simply, I have become convinced of immortality. We have talked of this, and I remember of telling you when we first knew each other that my idea of immortality was that we live afterward in the memory of those whose lives we have touched and in our accomplishments during life. And that has been enough for me. But now it is not. Whichever of us goes first (and it could be me, dear), *that* one will go on knowing about the other.

Rachel's response was gentle. As far as a continuation of consciousness beyond death, she was willing to dwell in mystery, and invoked the words of Swedish oceanographer Otto Pettersson, who, at the end of his long life, told his son that he would be sustained "by an infinite curiosity as to what was to follow." But there was one thing about which Carson did have some certainty, and that was in her sense of what she called "material immortality," where our bodies are first broken down by decay, then resurrected physically in new cellular arrangements. She summoned the words she had penned in an early piece, "Undersea," published in 1937 by the *Atlantic Monthly*.

> Individual elements are lost to view, only to reappear again and again in different incarnations in a kind of material immortality.... Against this cosmic background the life span of a particular plant or animal appears not as a drama complete in itself, but only as a brief interlude in a panorama of endless change.

She acknowledged to Dorothy that this is a "biologist's philosophy," and her observations are both spiritually beautiful and scientifically astute. It is a law of physics that all matter is conserved—our bodies return, return, return. This is the message of ecologists, and of mystics—that each life is radically connected to all of life, always, with nothing so small that it can be lost.

Rachel's thinking here always calls to mind Julian of Norwich, who, in the fourteenth century, was the first woman known to publish a book in the English language (and so a pioneer, as Rachel was in being a female science writer, uncommon in her time), and (again like Rachel) loved cats. In a famous vision, Julian saw "a small thing, the size of a hazelnut," nestled in the palm of her hand. "It was round as a ball." Confused, the saint pondered her tiny nut. Mirabai Starr translates Julian's epiphany:

I looked at it with the eyes of understanding and thought, *What can this be?* And the answer came to me: *It is all that is created.* I was amazed that it could continue to exist. It seemed to me to be so little that it was on the verge of dissolving into nothingness.

Yet she was comforted that this fragile thing was cradled within a limitless sacred whole. Julian delighted in her holy little hazelnut, presaging the modern view of earth as a "pale blue dot" given us by astronauts. It led, in part, to her complex proclamation from the beginning of this book: though all may be difficult, even despairing, and ever so, yet *All shall be well, and all shall be well, and all manner of thing shall be well.*

My grown-up vision of earthen grace remains influenced by a line from the creed I recited weekly as a child at Saint Anthony's Catholic Church in the mercurial green Pacific Northwest even before I discovered Frog Church: *I believe in the seen and the unseen.* I remember standing there like a tree amidst the pews, not knowing the science yet, but knowing something true.

The seen: My arms and hair flew into the air as branches, clothed with mosses and lichens, crawling with shiny black beetles, covered with birds, rustling in the slightest breeze, wet, sometimes, with rain, with dewfall.

And the unseen: Roots shot through the soles of my patent leather Mary Janes and into the dark earth, where they knit with other roots, rounded ancient stones, wove with worms and grubs and bodies of the furred and feathered returning to soil. All mingling mercilessly, gloriously, and always.

All shall be well, in whatever tangled, unknowable, difficult, beautiful way that wellness unfolds. Our lives are irrevocably entwined with this unfurling. Though we can't know exactly where we are going, or what will happen, still we journey together by choice and in grace, foot by foot, upon our troubled and beloved earth.

RETURN, RETURN, RETURN

Acknowledgments

I would like to begin by acknowledging that I live and work on the traditional land of the first people of Seattle, the Duwamish People past and present, and honor with gratitude this land and the Duwamish Tribe.

A book is a forest, and *Rooted* would not exist without the web of beings who brought it into the world. Thank you to my agent, Elizabeth Wales, for faith and vision. Thank you to my editor, Tracy Behar, for brilliance and humor and wisdom. Thank you to illustrator Helen Nicholson, for bringing such enchantment to these pages. Thank you to Ian Straus, and the whole extraordinary team at Little, Brown Spark. Thank you to the librarians of the Seattle Public Libraries, and all librarians and library folk everywhere—the true guardians of deep magic. Thank you to the many experts who lent their insight to this project.

In this uncertain time, I am more grateful than ever for my dear friends, family, and community. I offer my deepest thanks to the following odd tangle of humans for all they have shared with so much generosity of spirit: Trileigh Tucker (woodland conversation), Maria Dolan (commiseration), Lynda Mapes (trees, orcas, and unicorns), Chris Williams (working-artist happy hours), David Laskin (gardens and growing things), David Williams (the stability of stones), Gavin Van Horn (mischief), The Sirens (joyful dance), the Unspeakables (I could tell you why, but I'd have to kill you), the sisters of Saint Placid Priory (radical hospitality), Ginny Furtwangler (much-needed prayers), Al Furtwangler (Dr. Definition), Jill Story (sweet steadiness), Kelly Haupt (sisterhood and physics forever), Jerry Haupt (never-ending creativity), Irene Haupt (wise and loving common sense), Claire (awe and wonder that does not cease), and Tom (the evolution of love).

Selected Bibliography

Abram, David. "Magic and the Machine." *Emergence Magazine*, November 2018.

———. *The Spell of the Sensuous: Perception and Language in a More-than-Human World*. New York: Vintage, 1997.

Atchley, Ruth Ann, et al. "Creativity in the Wild: Improving Creative Reasoning through Immersion in Natural Settings." *PLOS OnE*, December 12, 2012.

Bashō, Matsuo. *The Narrow Road to the Deep North and Other Travel Sketches*. London: Penguin Classics, 1966.

Beckoff, Marc. *The Emotional Lives of Animals: A Leading Scientist Explores Animal Joy, Sorrow, and Empathy—and Why They Matter*. Novato, CA: New World Library, 2007.

Bent, A. C. *Life Histories of North American Birds*. New York: Dover, 1953.

Berry, Thomas. *The Dream of the Earth*. San Francisco: Sierra Club Books, 1988.

———. *The Great Work: Our Way into the Future*. New York: Bell Tower, 1999.

Blackie, Sharon. *Foxfire, Wolfkin, and Other Stories of Shape-Shifting Women*. Kent, UK: September Publishing, 2019.

Bogard, Paul. *The End of Night: Searching for Darkness in an Age of Artificial Light*. New York: Little, Brown, 2014.

Bowman, Katy. *Move Your DNA: Restore Your Health Through Natural Movement*. Sequim, WA: Propriometrics Press, 2017.

Carson, Rachel. *The Sense of Wonder*. New York: Harper & Row, 1965.

Chevalier, Gaétan, et al. "Earthing: Health Implications of Reconnecting the Human Body to the Earth's Surface Electrons." *Journal of Environmental and Public Health,* January 2012.

Clarke, John, trans. *Story of a Soul: The Autobiography of Saint Thérèse of Lisieux*. Washington, DC: ICS Publications, 1996.

Davis, Wade. *The Wayfinders: Why Ancient Wisdom Matters in the Modern World*. Toronto: House of Anansi Press, 2009.

Deignan, Kathleen, ed. *Thomas Merton: When the Trees Say Nothing—Writings on Nature*. Notre Dame, IN: Sorin Books, 2003.

Dosen, Stephanie. *Woodland Knits: Over Twenty Enchanting Patterns*. Newtown, CT: Taunton Press, 2013.

Ellsberg, Robert. "Egeria." *Give Us This Day,* April 2020.

Estés, Clarissa Pinkola. *Women Who Run with the Wolves*. New York: Ballantine Books, 1992.

Flinders, Carol Lee. *Enduring Grace: Living Portraits of Seven Women Mystics*. San Francisco: HarperSanFrancisco, 1993.

Francis of Assisi. "The Canticle of the Creatures." In Regis J. Armstrong, et al., eds., *Francis of Assisi: Early Documents*, vol. 1. New York: New City Press, 1999.

Francis (pope). *Laudato Si': On Care for Our Common Home*. Encyclical Letter. Huntington, IN: Our Sunday Visitor, 2015.

Franklin, Ralph William, ed. *The Poems of Emily Dickinson.* Cambridge, MA: Harvard University Press, 1998.

Freeman, Martha, ed. *Always, Rachel: The Letters of Rachel Carson and Dorothy Freeman, 1952–1964.* Boston: Beacon Press, 1995.

Furnass, B. "Health Values," in J. Messer and J. G. Mosley, eds., *The Value of National Parks to the Community: Values and Ways of Improving the Contribution of Australian National Parks to the Community.* Sydney: University of Sydney and Australian Conservation Foundation, 1979.

Gooley, Tristan. *The Lost Art of Reading Nature's Signs.* New York: Workman Publishing, 2014.

Görres, Ida Friederike. *The Hidden Face: A Study of Saint Thérèse of Lisieux.* San Francisco: Ignatius Press, 2003.

Gregoire, Carolyn. "The New Science of the Creative Brain on Nature." *Outside Online,* March 18, 2016.

Griffin, Susan. *Woman and Nature: The Roaring Inside Her.* New York: Harper & Row, 1978.

Griffiths, Jay. *A Country Called Childhood: Children and the Exuberant World.* Berkeley, CA: Counterpoint, 2014.

———. "Forests of the Mind." *Aeon,* October 12, 2012.

Habegger, Alfred. *My Wars Are Laid Away in Books: The Life of Emily Dickinson.* New York: Random House, 2001.

Haile, Rahawa. "Forest Bathing: How Microdosing on Nature Can Help with Stress." *The Atlantic,* June 30, 2017.

Halifax, Joan. *The Fruitful Darkness: Reconnecting with the Body of the Earth.* San Francisco: HarperSanFrancisco, 1993.

Harjo, Joy. *She Had Some Horses: Poems.* New York: W. W. Norton, 2008.

Hart, Patrick, and Jonathan Montaldo, eds. *The Intimate Merton: His Life from His Journals*. San Francisco: HarperSanFrancisco, 1999.

Haskell, David George. *The Songs of Trees*. New York: Viking, 2017.

Hirshfield, Jane. *Come, Thief.* New York: Alfred A. Knopf, 2011.

Hoff, Benjamin. *The Singing Creek Where the Willows Grow: The Mystical Nature Diary of Opal Whiteley*. New York: Penguin Books, 1986.

Johnson, Thomas H., ed. *Emily Dickinson: Selected Letters.* Cambridge, MA: Harvard University Press, 1958.

Kaza, Stephanie. "Walking into the Unknown." In Lisa Bach, ed., *365 Travel: A Daily Book of Journeys, Meditations, and Adventure*. Palo Alto: Travelers' Tales, 2001.

Kimmerer, Robin Wall. "Speaking of Nature." *Orion*, March/April 2017.

Le Guin, Ursula K. *The Books of Earthsea: The Complete Illustrated Edition*. New York: Saga Press, 2018.

LeMay, Kristin. *I Told My Soul to Sing: Finding God with Emily Dickinson*. Brewster, MA: Paraclete Press, 2013.

Louv, Richard. *Our Wild Calling: How Connecting with Animals Can Transform Our Lives—and Save Theirs*. Chapel Hill, NC: Algonquin Books, 2019.

Macfarlane, Robert. *Landmarks*. New York: Penguin Books, 2016.

Macfarlane, Robert, and Jackie Morris. *The Lost Words: A Spell Book*. London: Hamish Hamilton, 2018.

Macy, Joanna. *World as Lover, World as Self*. Berkeley, CA: Parallax Press, 1991.

Macy, Joanna, and Molly Young Brown. *Coming Back to Life.* Gabriola Island, BC: New Society Publishers, 2014.

Maitland, Sara. *From the Forest: A Search for the Hidden Roots of Our Fairy Tales.* Berkeley, CA: Counterpoint, 2012.

Maller, Cecily Jane, et al. "Healthy Parks, Healthy People: The Health Benefits of Contact with Nature in a Park Context." Parks Stewardship Forum, January 2009.

Mapes, Lynda. *Orca: Shared Waters, Shared Home.* Seattle: Mountaineers Books, 2021.

———. *Witness Tree: Seasons of Change with a Century-Old Oak.* Seattle: University of Washington Press, 2019.

Maskowitz, David. *Wildlife of the Pacific Northwest.* London: Timber Press, 2010.

McDougall, Christopher. *Born to Run: A Hidden Tribe, Super-athletes, and the Greatest Race the World Has Never Seen.* New York: Alfred A. Knopf, 2009.

Miyazaki, Yoshifumi, et al., eds. "Shinrin-Yoku (Forest Bathing) and Nature Therapy: A State-of-the-Art Review." *International Journal of Environmental Research and Public Health,* July 28, 2017.

Murphy, Francis, ed. *Walt Whitman: The Complete Poems.* New York: Penguin Classics, 2005.

Nhat Hanh, Thich. *The Miracle of Mindfulness.* Boston: Beacon Press, 1999. First published 1976.

Ober, Clinton, et al. *Earthing: The Most Important Health Discovery Ever!* Laguna Beach, CA: Basic Health, 2014.

O'Donohue, John. *Anam Cara: A Book of Celtic Wisdom.* New York: HarperCollins, 1997.

————. *Beauty: The Invisible Embrace*. New York: HarperCollins, 2005.

Oliver, Mary. *New and Selected Poems*. Boston: Beacon Press, 1992.

Plotkin, Bill. *Nature and the Human Soul: Cultivating Wholeness and Community in a Fragmented World*. Novato, CA: New World Library, 2008.

————. *Soulcraft: Crossing into the Mysteries of Nature and Psyche*. Novato, CA: New World Library, 2003.

Pojar, Jim, and Andrew MacKinnon. *Plants of the Pacific Northwest Coast*. Redmond, WA: Lone Pine, 2016.

Pollan, Michael. "The Intelligent Plant." *The New Yorker*, December 16, 2013.

Powers, Richard. *The Overstory*. New York: W. W. Norton, 2018.

Raichle, Marcus E., et al. "A Default Mode of Brain Function." *Proceedings of the National Academy of Sciences*, vol. 98, no. 2 (2001).

Revkin, Andrew. "Confronting the 'Anthropocene.'" *Dot Earth: New York Times Blog*, May 11, 2011.

Sheldrake, Merlin. *Entangled Life: How Fungi Make Our Worlds, Change Our Minds, and Shape Our Futures*. New York: Random House, 2020.

Simard, Suzanne, et al. "How Trees Talk to Each Other." TED Talks, 2016.

————. "Mycorrhizal Networks: Mechanisms, Ecology, and Modelling." *Fungal Biology Reviews*, April 2012.

————. "Net Transfer of Carbon Between Ectomycorrhizal Tree Species in the Field." *Nature*, July 1997.

Smith, Jen Rose. "Sleeping Alone in the Woods While Female." *Outside Online*, December 20, 2016.

Solnit, Rebecca. *Hope in the Dark: Untold Histories, Wild Possibilities*. Chicago: Haymarket Books, 2016.

———. "Our Storied Future: Never Underestimate the Power of an Idea." *Orion*, January/February 2008.

———. *Wanderlust: A History of Walking*. New York: Penguin Books, 2001.

Stamets, Paul. *Mycelium Running: How Mushrooms Can Help Save the World*. Berkeley, CA: Ten Speed Press, 2005.

Starr, Mirabai, trans. *The Showings of Julian of Norwich: A New Translation*. Charlottesville, VA: Hampton Roads, 2013.

———. *Teresa of Ávila: The Book of My Life*. Boston: New Seeds Books, 2007.

Subramanian, Meera. "Anthropocene Now: Influential Panel Votes to Recognize Earth's New Epoch." *Nature*, May 21, 2019.

Tetlow, Adam. *Designa: Technical Secrets of the Traditional Visual Arts*. New York: Bloomsbury, 2014.

Turner, Jack. *The Abstract Wild*. Tucson: University of Arizona Press, 1996.

Tweit, Susan. *Bless the Birds*. Berkeley, CA: She Writes Press, 2021.

Villanueva, Edgar. *Decolonizing Wealth: Indigenous Wisdom to Heal Divides and Restore Balance*. Oakland, CA: Berrett-Koehler Publishers, 2016.

Williams, Florence. *The Nature Fix: Why Nature Makes Us Happier, Healthier, and More Creative*. New York: W. W. Norton, 2017.

Windling, Terri. "The Speech of Animals." *Myth & Moor*, February 10, 2015.

Young, Jon, et al. *Coyote's Guide to Connecting with Nature*. Shelton: WA: Owlink Media, 2008.

Copyright Acknowledgments

About the Author

Lyanda Lynn Haupt is a naturalist, ecophilosopher, and author of several books, including *Mozart's Starling, The Urban Bestiary, Crow Planet, Pilgrim on the Great Bird Continent,* and *Rare Encounters with Ordinary Birds*. A winner of the Washington State Book Award and the Sigurd F. Olson Nature Writing Award, she lives in Seattle with her husband and daughter.